Lecture Notes in Control and Information Sciences

Edited by A. V. Balakrishnan and M. Thoma

Vol. 1: Distributed Parameter Systems: Modelling
and Identification
Proceedings of the IFIP Working Conference,
Rome, Italy, June 21–26, 1976
Edited by A. Ruberti
V, 458 pages. 1978

Vol. 2: New Trends in Systems Analysis
International Symposium, Versailles,
December 13–17, 1976
Edited by A. Bensoussan and J. L. Lions
VII, 759 pages. 1977

Vol. 3: Differential Games and Applications
Proceedings of a Workshop, Enschede, Netherlands,
March 16–25, 1977
Edited by P. Hagedorn, H. W. Knobloch, and G. J. Olsder
XII, 236 pages. 1977

Vol. 4: M. A. Crane, A. J. Lemoine
An Introduction to the Regenerative Method for
Simulation Analysis
VII, 111 pages. 1977

Vol. 5: David J. Clements, Brian D. O. Anderson
Singular Optimal Control: The Linear Quadratic Problem
V, 93 pages. 1978

Vol. 6: Optimization Techniques
Proceedings of the 8th IFIP Conference on Optimi-
zation Techniques, Würzburg, September 5–9, 1977
Part 1
Edited by J. Stoer
XIII, 528 pages. 1978

Vol. 7: Optimization Techniques
Proceedings of the 8th IFIP Conference on Optimi-
zation Techniques, Würzburg, September 5–9, 1977
Part 2
Edited by J. Stoer
XIII, 512 pages. 1978

Vol. 8: R. F. Curtain, A. J. Pritchard
Infinite Dimensional Linear Systems Theory
VII, 298 pages. 1978

Vol. 9: Y. M. El-Fattah, C. Foulard
Learning Systems:
Decision, Simulation, and Control
VII, 119 pages. 1978

Vol. 10: J. M. Maciejowski
The Modelling of Systems with Small Observation Sets
VII, 241 pages. 1978

Vol. 11: Y. Sewaragi, T. Soeda, S. Omatu
Modelling, Estimation, and Their Applications for
Distributed Parameter Systems
VI, 269 pages. 1978

Vol. 12: I. Postlethwaite, A. G. J. McFarlane
A Complex Variable Approach to the Analysis of
Linear Multivariable Feedback Systems
IV, 177 pages. 1979

Vol. 13: E. D. Sontag
Polynomial Response Maps
VIII, 168 pages. 1979

Vol. 14: International Symposium on Systems
Optimization and Analysis
Rocquentcourt, December 11–13, 1978;
IRIA LABORIA
Edited by A. Bensoussan and J. Lions
VIII, 332 pages. 1979

Vol. 15: Semi-Infinite Programming
Proceedings of a Workshop, Bad Honnef,
August 30 – September 1, 1978
V, 180 pages. 1979

Vol. 16: Stochastic Control Theory
and Stochastic Differential Systems
Proceedings of a Workshop of the „Sonder-
forschungsbereich 72 der Deutschen Forschungs-
gemeinschaft an der Universität Bonn"
which took place in January 1979 at Bad Honnef
VIII, 615 pages. 1979

Vol. 17: O. I. Franksen, P. Falster, F. J. Evans
Qualitative Aspects of Large Scale Systems
Developing Design Rules Using APL
XII, 119 pages. 1979

Vol. 18: Modelling and Optimization of Complex
Systems
Proceedings of the IFIP-TC 7 Working Conference
Novosibirsk, USSR, 3–9 July, 1978
Edited by G. I. Marchuk
VI, 293 pages. 1979

Vol. 19: Global and Large Scale System Models
Proceedings of the Center for Advanced Studies (CAS)
International Summer Seminar
Dubrovnik, Yugoslavia, August 21–26, 1978
Edited by B. Lazarević
VIII, 232 pages. 1979

Vol. 20: B. Egardt
Stability of Adaptive Controllers
V, 158 pages. 1979

Vol. 21: Martin B. Zarrop
Optimal Experiment Design for
Dynamic System Identification
X, 197 pages. 1979

Lecture Notes in Control and Information Sciences

Edited by A.V. Balakrishnan and M. Thoma

21

Martin B. Zarrop

Optimal Experiment Design for Dynamic System Identification

Springer-Verlag Berlin Heidelberg GmbH 1979

Author
Dr. Martin B. Zarrop
Control Systems Centre
University of Manchester Institute of Science and Technology
Sackville Street
Manchester M60 1QD

ISBN 978-3-540-09841-6   ISBN 978-3-540-38993-4 (eBook)
DOI 10.1007/978-3-540-38993-4

ABSTRACT

This work is concerned with the problem of experiment design for
the efficient identification of a linear single input, single output
dynamic system from input-output data in the presence of disturbances.
The experimenter is allowed to select certain factors under his control
(input signal, output filter, sampling times), subject to suitable
constraints, in order to maximise information from an experiment. A
frequency domain approach to the test signal/sampling rate design problem
is adopted and the cost criterion is chosen to be a suitable convex
scalar function of the inverse Fisher information matrix.

A geometrical approach to the design problem is developed for
both continuous-time and discrete-time systems, based on the theory of
Tchebycheff systems and their associated moment spaces. Conditions are
derived for the existence of certain minimal representations of the optimal
input spectrum, leading to a reduction in the dimension of the design
optimization problem. In particular, for a restricted class of model
structures, the optimal input spectrum contains the minimum number of
frequencies consistent with a persistently exciting signal.

A class of sequential design algorithms is proposed with proven
global convergence to a D-optimal design. Comparison is made of the
computational efficiency of a number of these algorithms.

ACKNOWLEDGEMENTS

I would like to thank my supervisor Professor David Q. Mayne for his valuable guidance and great patience. I am also deeply grateful to Dr. Graham C. Goodwin who inspired this research and gave constant encouragement from afar.

My special thanks go to Robin Becker for many helpful discussions and for his programs which were used for the computational parts of this work. I am also grateful to the staff and students of the Control Section at Imperial College for their many useful suggestions and discussions, particularly Dr. R.B. Vinter, Mr. H.H. Johnson, Dr. R.L. Payne and Hossein Javaherian.

Finally, I wish to express my thanks to the Science Research Council for financial support and to Linden Rice for her excellent typing.

CONTENTS

Page

CONVENTIONS AND SYMBOLS

The system of numbering and cross-referencing is standard: within each section, equations, theorems etc. are given a single number and only this number is given when reference is made from within the same section. When reference is made from another section of the same chapter, the section number is also given. A similar convention applies to the numbering of sections within chapters.

The end of a proof or of a particular train of thought is denoted by #.

An asterisk prefixes a theorem or result considered original by the author, e.g. *Theorem 1.

The usage of other commonly used symbols is given below, together with the section in which the symbol is introduced.

Symbol		Section
a_i	coefficient of polynomial A	2.2
$A(\cdot)$	denominator polynomial of system transfer function (t.f.)	2.2
b_i	coefficient of polynomial B	2.2
$B(\cdot)$	numerator polynomial of system t.f.	2.2
c_i	coefficient of polynomial C	2.2
$C(\cdot)$	denominator polynomial of noise t.f.	2.2
d_i	coefficient of polynomial D	2.2
$d(\cdot,\cdot)$	generalised variance	4.4
$D(\cdot)$	numerator polynomial of noise t.f.	2.2
$\mathcal{D}_1$	set of discrete normalised design measures	2.7
e_i	white noise sequence	1.3

E	expectation operator	2.3
$h(\cdot)$	$u \rightarrow \partial\varepsilon/\partial\theta$	2.6
$I(\cdot)$	design index	2.8
k	discrete time	2.2
L	log likelihood function	2.3
m	degree of B	2.2
M	information matrix	2.3
$\mathcal{M}$	set of information matrices	2.7
$M_c^{(p)}$	cone in R^p	3.5
n	degree of A	2.2
N	number of data points	1.3
p	number of system parameters	2.4
q	degree of C	2.2
r	degree of D	2.2
s	Laplace transform variable	5.2
$S(\cdot)$	feasible set	4.8
t	continuous time	5.2
T	experiment time	5.3
u	system input	2.3
v	T-system function	3.4
V	T-system determinant	3.4
W	cost weighting matrix	2.1
y	system output	1.3
z	unit forward shift operator	2.2
δ	Dirac delta function	2.6
δ_{ij}	Kronecker delta	2.4
Δ	sampling interval	6.4
ε_i	residual sequence	2.4

Chapter 1

PRELIMINARIES

1.1   INTRODUCTION

This work is concerned with the planning of experiments for

efficient identification of linear dynamic systems.

Practically, the experimenter has to decide on the purpose of the

experiment, which variables to measure and how to measure them.   These

decisions will be taken under various physical, technological and

economic constraints and conditioned by the amount of prior information

available.   Indeed, the absense of prior knowledge of the system under

investigation renders the experiment design problem meaningless.   In

this thesis it is assumed that the experimenter has only the form of the

input signal and/or the output sampling rate left to determine in order

to minimise uncertainty in a finite set of parametric constants.   These

parameters completely determine the process characteristics.   Subject to

certain constraints the control variables are chosen to optimise a

suitable measure of goodness of the experiment.

Conceptually, this aspect of dynamic system identification is an

extension of the experiment design problem on which statisticians have

written extensively over the past sixty years.   In the present work

certain facets of the design problem are investigated from a geometric

standpoint drawing on the classical theory of Tchebycheff systems and

their associated moment spaces.

Applications of the results are clearly most useful in areas where

experimentation is expensive, such as aircraft flight tests and production

line tests.   However, the problem of characterising optimal experiment

designs is an absorbing field of theoretical investigation in its own
right.

1.2   IDENTIFICATION AND EXPERIMENT DESIGN

The most important stages in the overall identification procedure
are the choice of experimental goal and the collection of prior
knowledge. The place of experiment design in this procedure is shown
schematically in figure 1. Efficient design of experiments can only
begin when the gross characteristics have been determined, if necessary
by a number of preliminary experiments. In this thesis it is assumed
that the model structure is known and that preliminary information
concerning the process parameters is embodied in a prior probability
distribution, sharply peaked at some value close to that of the true
parameter vector. More general Bayesian approaches [A1] [P1] [L1],
involving less restrictive assumptions on the prior distribution, are
analytically intractable and computationally prohibitive except in the
most trivial cases.

Specific identification methods and theoretical aspects of
identification problems have already been extensively surveyed in the
literature, e.g. Eykhoff et al [E1] [E2], Astrom and Eykhoff [A2],
Nieman et al [N1] and the two textbooks by Graupe [G1] and Eykhoff [E3].
The reader is referred to these sources for details.

Surveys of applications of identification techniques to real
processes have been carried out by Rault [R1], Baeyens and Jaquet [B1],
Rajbman [R2] and Gustavsson [G2]. This last paper presents general
principles governing experiment design for identification, e.g. choice
of test signal, sampling rate, experiment time, etc.

1.2

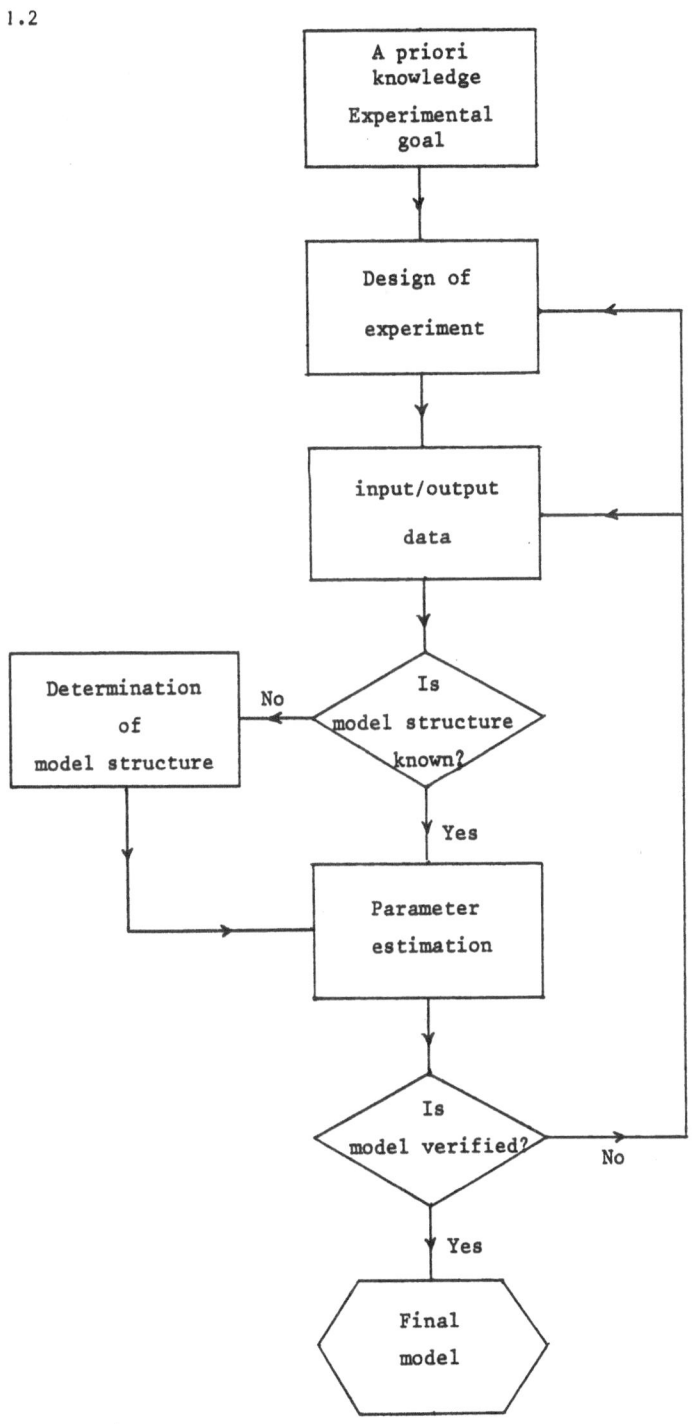

Figure 1 :   The identification process

4

The aspects of experiment design relevant here, i.e. input signal
design and choice of sampling intervals, are discussed in the next two
sections.

1.3 OPTIMAL INPUT DESIGN

The design of optimal inputs for static systems is well covered in
the statistical literature. In particular, the book by Fedorov [F1]
and the survey paper by St. John and Draper [S1] together provide a
comprehensive discussion of the static design problem. Much of the
present research in this area is based on the approach developed since
1959 by Kiefer and Wolfowitz [K1]-[K6], Karlin and Studden [K7], Whittle
[W1] and others.

A linear (in the parameters) multiple input single output static
system is described by equation (1):

$$y_j = \theta^T f(x_j) + e_j, \qquad j = 1, 2, \ldots, N \qquad (1)$$

where y_j is the jth observation and $\{e_j\}_1^N$ is a sequence of uncorrelated
and identically distributed random variables with zero mean and variance
σ^2. The px1 vector $f(\cdot)$ is assumed known and continuous on some compact
set χ (experimental region).

From the observations $\{y_j\}_1^N$ a minimum variance unbiased linear
estimator, $\hat{\theta}$, of the px1 parameter vector θ can be derived by the least
squares procedure. The covariance matrix of $\hat{\theta}$ is given by:

$$\text{cov } \hat{\theta} = \sigma^2 [\sum_{j=1}^N f(x_j) f^T(x_j)]^{-1} \qquad (2)$$

The design problem consists of selecting vectors x_i, $i = 1, \ldots, N$ from

χ such that the design defined by these N vectors is, in some specified

sense, optimal. This is usually carried out by choosing the design to

minimise some chosen scalar function of cov $\hat{\theta}$.

Kiefer and Wolfowitz [K2] extend this concept of design by introducing

a measure ξ on χ. It is then possible to demonstrate the equivalence of

certain optimality criteria. This equivalence theory has been used to

construct sequential design procedures converging on optimal designs [W2]

[F2]. The extensions of this work to multiple output systems [F3] and

models that are nonlinear in the parameters [W3] can be carried out in a

straightforward manner.

The literature dealing with the design of optimal inputs for dynamic

systems is more sparse and has mainly taken a time-domain approach. In

particular:

Levin (1960, [L2]) employs a Markov estimator for the parameters

in a time invariant weighting function model of a linear system.

In the case of white output noise, he shows that an input with

impulsive autocorrelation is optimal with respect to several

optimality criteria.

Levadi (1966, [L3]) considers a linear time varying system with

non-stationary coloured output noise. The time variations are

assumed completely known and the system output is assumed linear

in the parameters. A Markov estimator is used and necessary

conditions for optimality are derived.

The exact covariance of a particular estimator is very difficult

to obtain. Many authors assume the existence of an asymptotically

efficient parameter estimator (e.g. maximum likelihood) so that the parameter covariance matrix is approximated by the inverse of the Fisher

information matrix M (Cramer Rao lower bound) for long data lengths. The

input design problem can then be formulated in control theoretic terms

with the input chosen to minimise some suitable scalar function of M.

Nahi and Wallis (1969, [N2]) employ a control theoretic approach

in design of optimal inputs for nonlinear single input single output

(SISO) systems with white observation noise. The choice of cost function

is trace WM where W is a constant weighting matrix.

Aoki and Staley [A2][A3], Nahi and Napjus [N3][N4] and Mehra [M1]

also use optimality criteria of the form trace WM. This criterion leads

to a quadratic optimization problem which is easy to solve, but may

produce optimal inputs which are not persistently exciting [A4][L4].

This has been pointed out by Zarrop and Goodwin [Z1][G3], Reid [R3],

Tse [T1] and Mehra [M2] and is further discussed in Chapter 2 of this

work.

Viort (1972, [V1]) was the first to make the connection between the

input design problem for dynamic systems and the experiment design problem

for static systems by employing a frequency domain approach. His analysis

is restricted to SISO autoregressive-moving average (ARMA) models.

Goodwin and Payne (1973, [G4]; 1974, [P2]) and Van den Bos (1973,

[V2]) show that, for long record lengths, the optimal input signal for

linear SISO system parameter estimation can be characterised by its

spectral properties.

Mehra (1973, [M3]; 1974,[M4]) extends the work of Viort in applying

static experiment design theory to optimal parameter estimation in linear

MIMO systems, with energy or power constrained inputs. The set of average

information matrices $\bar{M}$ (i.e. average per sample, $\bar{M}_{ij} = \lim_{N \to \infty} \frac{1}{N} M_{ij}$)

corresponding to normalised input power is shown to be convex. Also any

$\bar{M}$ can be generated by an input with spectrum comprising at most $p(p+1)/2 +1$ discrete frequencies, where p is the number of unknown parameters.

Mehra proposes a sequential design procedure based on that of Fedorov [F2] converging to a globally optimal design. The main disadvantage of this algorithm is that it converges in an infinite number of steps and 'rounding-off' procedures are usually employed [F1][S1][G5].

Payne and Goodwin (1974, [P3]) examine the detailed structure of $\bar{M}$ for a general linear continuous-time SISO model with parametrically disjoint system and noise transfer functions. It is seen that $\bar{M}$ can be represented as a point in a p-dimensional Euclidean space where p is the number of unknown system parameters. This leads to an extension of Mehra's result by reducing the search for the optimal design to 2p-1 variables (i.e. p frequencies and p power proportions summing to unity).

Javaherian (1974, [J1]) computes D-optimal designs (i.e. optimality criterion is det $\bar{M}^{-1}$) for linear discrete time SISO models using the Fletcher-Powell conjugate gradient algorithm. This allows the number of frequencies in the input spectrum to be arbitrarily fixed but several different starting designs may be necessary to achieve global optimality as the algorithm only seeks local minima. The computer results indicate that in many cases optimal designs can be characterised by [(p+1)/2] frequencies, i.e. the smallest number consistent with a persistently exciting input signal [L4].

A more comprehensive survey of literature on the time-domain synthesis of optimal inputs is given by Mehra (1974, [M2]) and this paper contains material previously available only in technical reports [M3][M4].

1.4 OPTIMAL DESIGN OF SAMPLING RATE

The choice of sampling intervals also has a significant bearing on identification accuracy [G2].

Astrom (1968, [A5]) discusses uniform sampling of a simple Gauss-Markov process and shows that an optimal sampling rate exists in the sense of minimizing the variance of a parameter estimator.

Zarrop (1973, [Z2]) introduces an input term into Astrom's model and shows that the existence of a well-defined optimal sampling rate is dependent on the input amplitude constraint.

Gustavsson (1971, [G6]) calculates D-optimal uniform sampling rates for linear continuous-time SISO systems with white noise input and special structure. In general it is found that, where optima exist, a sampling rate corresponding to a Nyquist frequency twice to five times the highest breakpoint frequency or resonance frequency is reasonable.

Payne, Goodwin and Zarrop (1974, [P4]) consider the problem of joint determination of input spectra and sampling rate. It is shown that the introduction of a suitable filter allows the coupled design to be carried out in the frequency domain applying the approach developed by Mehra [M3].

Goodwin, Zarrop and Payne (1974, [G7]) propose a general approach to the joint design of presampling filter, nonuniform sampling rate and input signal. A suboptimal sequential procedure is used and a simple example indicates that nonuniform sampling can lead to substantial improvements in estimation accuracy.

Goodwin and Payne (1974, [G9]) provide a useful survey of current work on the sampling rate problem that includes material on the (suboptimal) sequential design of nonuniform sampling intervals and test signals.

Ng and Goodwin (1976, [N5]) show that a constant sampling rate strategy can be optimised by decomposing an experiment into at most a finite number of subexperiments each with its own constant sampling rate. The design scheme is extended to deal with diffuse prior distributions for the parameters.

1.5   REVIEW AND ORIGINAL CONTRIBUTIONS

In this work two main approaches are made to the experiment design problem as it arises in the identification of linear dynamic SISO systems. These approaches are introduced in Chapters 3 and 4 which contain the central original ideas and form the core of the work. The theoretical development of the basic concepts presented in these chapters has been the primary concern of the author and, to this end, relatively little emphasis has been placed on the practical implications of the results obtained.

First, the test signal/sampling rate design problem is analysed within a geometric framework based on classical Tchebycheff system theory [K8]. The basic ideas are presented and developed in Chapter 3 for a discrete time system with p estimable parameters and supported by the Main Appendix which brings together some established results from T-system theory for ease of reference. Secondly, the problem of attaining a D-optimal input by sequential design procedures is discussed in Chapter 4. The detailed survey below shows how the key chapters fit in and indicates the main results that the author believes to be original.

In Chapter 2 the basic elements of the input design problem are brought together, i.e. model structure, input constraints and optimality criteria (Sections 2, 7, 9). Central to the problem formulation is the calculation of the Fisher information matrix (Section 4) and the introduction of a frequency domain approach (Section 6) leading to the concept of an input design measure ξ. Minimal properties of test signals are discussed briefly (Section 8) and a necessary and sufficient condition for local identifiability of parameters is established [Theorem (2.8.2)], extending a result due to Mehra [M3]. It is shown that the information matrix $\bar{M}(\xi)$ is nonsingular iff a suitably defined design index $I(\xi)$ is not less than $p/2$ corresponding to a persistently exciting test signal.

It is also shown that identifiability problems can occur if the optimality criterion employed is linear in $\bar{M}$, e.g. trace $(\bar{W}\bar{M})$ [Theorem (2.9.1)]. For this class of criteria, optimality can be achieved with a single frequency design. This extends a result published by the author and G.C. Goodwin in IEEE Transactions on Automatic Control in 1975 [Z1].

A geometrical approach to the analysis of M, the set of information matrices corresponding to normalised input power, is developed in Chapter 3 (Sections 2 and 3). This is directed towards establishing conditions for the existence of optimal designs with low design index, thus reducing the dimension of the design problem. The theory of Tchebycheff systems is found to furnish a fruitful approach (Sections 4 and 5) leading to the concept of canonical designs with indices not exceeding $(p+1)/2$ and principal designs with index $p/2$ (Section 6). The T-system approach to optimal input design for dynamic systems is new and most of the material presented from Section 4 onwards is original. The crucial result, establishing the existence of a T-system, is Theorem (3.4.1).

The set M of information matrices is isomorphic to $M^{(p)}$, a *moment space* induced in R^p by the T-system. Geometrical considerations lead to sufficient conditions for the existence of optimal canonical designs, placing only weak restrictions on the optimality criterion (Sections 6 and 7). Some illustrative examples are presented in Appendix A.

An important case of interest occurs when $M^{(p)}$ is a hyperplane in R^p. Theorem (3.5.2) proves that this is equivalent to the process structure satisfying the conditions: $q = 0$, $m \geq n+r$ and Theorem (3.5.3) gives an alternative equivalent condition. In the hyperplane case, for any frequency ω_0, each nonsingular information matrix in M (including the optimum) can be generated by a unique canonical design whose spectrum includes ω_0 [Theorem (3.6.4)].

If $M^{(p)}$ is not a hyperplane, then Theorem (3.7.1) proves that any optimal design measure ξ^* must be discrete and that

$$I(\xi^*) \leq \tfrac{1}{2} \max \ (p+q-1, 2n+r) \qquad\qquad (*)$$

This upper bound can lead to a design problem of reduced dimension. In general, however, the number of frequencies in the optimal input spectrum may be equal to p if the hyperplane conditions are not satisfied [Appendix B]. In addition, the theorem implies that there can exist no optimal input with a continuous spectrum (e.g. white noise) unless $M^{(p)}$ is a hyperplane [Corollary (3.7.1)].

In Section 8, the case of normalised output power is analysed by introducing a suitable T-system and its associated moment space. Parallel results are obtained to those corresponding to normalised input power. In particular, the hyperplane conditions become $q = 0 = r$, thus placing no restriction on the system polynomials A, B [Theorem (3.8.1)].

In Chapter 4 attention is confined to designs that are optimal under the determinant criterion. It is shown that, if principal D-optimal designs exist, their power proportions have the values 2/p or 1/p [Theorem (4.2.1)] and this extends a result from static design theory [K8, p. 332]. In some cases, the optimal design can be completely determined analytically [Theorem (4.3.1)].

The main sections of Chapter 4 are concerned with possible
sequential design procedures converging to a D-optimal design (Sections
5, 7, 9). An extension to the Kiefer-Wolfowitz Equivalence Theorem [F1]
is proved [Theorem (4.8.1)] and this leads to Theorem (4.9.1), the
global convergence of a particular type of sequential design procedure
(S-algorithm) that embraces a number of algorithms previously proposed
by other authors [F1][M3][A6]. The rates of convergence of some of these
algorithms are compared in an appendix.

The sequential design approach leads to some further properties
of D-optimal designs (Section 6). In particular, for normalised input
power, no input frequency in a D-optimal design can contain power
exceeding $2/p$. This is an extension to the dynamic case of a result due
to Atwood [A6].

In Section 10, a *sequential* frequency 'rounding-off' procedure is
proposed that has only a second-order effect on the information matrix
at each iteration. In practice, this procedure speeds up convergence
of the sequential design algorithms to a final design with sparse
spectrum.

The material presented in Chapters 5 and 6 is new.

Chapter 5 extends the T-system approach to input design developed in
Chapter 3 for the discrete-time case to continuous-time systems with p
estimable parameters. Many of the previous results can be carried over
with little modification provided that high spectral frequencies are
treated with care (Sections 4 and 6). In the case when the system and
noise transfer functions have the same number of infinite zeros, i.e.
$n-m = q-r$ (Section 8) the infinite frequency limit can be taken and the
finite-interval theory applied unchanged. In an appendix, the problem
of using optimal designs based on approximate parameter estimates is

discussed using a simple example.

Chapter 6 uses the framework developed in Chapter 5 as a basis for analysis of a joint test signal/sampling rate scheme first proposed in the experiment design context by Payne [P2]. Geometrical difficulties arise in that the information matrix set is not necessarily either closed [Result (6.5.2)] or convex [Result (6.5.3)]. This corrects a result stated by Goodwin and Payne [G9]. Theorem (6.5.1) proves, however, that it is sufficient to consider only designs with discrete spectra whose design indices do not exceed p+2. A suitable design algorithm is then formulated (Section 6) and some properties of optimal designs are discussed based on certain interlacing properties of canonical design spectra (Section 7).

Finally, Chapter 7 contains some brief concluding remarks and pointers to further research.

Chapter 2

PROBLEM STATEMENT

2.1 INTRODUCTION

In this chapter the basic elements of the input design problem are brought together, i.e. model structure, input constraints and optimality criteria (Sections 2, 7, 9). Central to the problem formulation is the calculation of the Fisher information matrix (Section 4) and the introduction of a frequency domain approach (Section 6). Minimal properties of test signals are discussed briefly (Section 8) and a necessary and sufficient condition for local identifiability is proved. It is shown that the choice of a certain class of optimality criteria (including $\{\text{trace } (\overline{WM})\}^{-1}$) can lead to identifiability problems.

2.2 MODEL STRUCTURE

The model considered represents a linear time invariant discrete time SISO system with input sequence $\{u_k\}$ and output sequence $\{y_k\}$. The output is assumed to be corrupted by noise having rational spectral density $\Psi(z)$ with no poles or zeros on the unit circle. It can be shown (see e.g. Astrom [A7, p. 98ff]) that there exists a factorization of $\Psi(z)$ such that

$$\Psi(z) = H(z)H(z^{-1}) \tag{1}$$

where

$$H(z) = D(z)/C(z) \tag{2}$$

2.2

5

and C and D are polynomials whose zeros lie outside the closed unit

disc. This results in a noise model

$$y_k = H(z^{-1})e_k \tag{3}$$

where $\{e_k\}$ is a sequence of independent zero mean unit variance random

variables and z is the unit forward shift operator (i.e. $zx_k = x_{k+1}$).

The complete model can be written

$$y_k = Z(z^{-1})u_k + H(z^{-1})e_k \tag{4}$$

where Z is the rational transfer function u → y. It is assumed that

the model represents a stable, minimum-phase system with time delay

s so that

$$Z(z) = z^s B(z)/A(z) \tag{5}$$

where A and B are polynomials whose zeros lie outside the closed unit

disc.

The final model can therefore be written

$$y_k = z^{-s} \frac{B(z^{-1})}{A(z^{-1})} u_k + \frac{D(z^{-1})}{C(z^{-1})} e_k \tag{6}$$

where

$$A(z) = 1 + a_1 z + \ldots + a_n z^n \tag{7a}$$

$$B(z) = b_0 + b_1 z + \ldots + b_m z^m \quad (b_0 \neq 0) \tag{7b}$$

$$C(z) = 1 + c_1 z + \ldots + c_q z^q \qquad\qquad (7c.)$$

$$D(z) = d_0 + d_1 z + \ldots + d_r z^r \qquad (d_0 \neq 0) \qquad\qquad (7d)$$

It is assumed that the polynomials A, B, C, D are relatively prime unless otherwise stated and that n, m, q, r and s are known integers. The vector of process parameters

$$\beta = (a_1, \ldots, a_n, b_0, \ldots, b_m, c_1, \ldots, c_q, d_0, \ldots, d_r)^T \in R^{p'}$$

$$p' = m + n + q + r + 2$$

is to be estimated from input-output data.

2.3 CRAMER-RAO LOWER BOUND

Accuracy of estimation can be assessed by forming the covariance matrix of the parameter estimator. This is difficult to do in general but considerable simplification is achieved for a wide class of estimators.

Let y denote the vector containing the first N output values as components, i.e.

$$y \triangleq (y_1, y_2, \ldots, y_N)^T \in Y \in R^N$$

where Y denotes the allowable set of such vectors. Similarly

$$u \triangleq (u_1, \ldots, u_N)^T \in U \in R^N$$

where U denotes the allowable set of input vectors.

The likelihood function is defined to be $p(y|\beta,u)$, the conditional probability density of y given β and u and the log-likelihood function $L(y,u,\beta)$ is defined by

$$L(y,u,\cdot) = \log p(y|\cdot,u) \qquad (1)$$

and regarded as a function of β.

Then the Cramer-Rao inequality states that, subject to weak regularity conditions (see Silvey [S2, pp. 35-37]), the covariance matrix of any unbiased estimator $\hat{\beta}$ of β satisfies

$$\text{cov } \hat{\beta} \geq M_\beta^{-1} \qquad (2)$$

where M_β is the Fisher information matrix defined by

$$M_\beta = E_{y|\beta,u}\left[\left(\frac{\partial L(y,u,\beta)}{\partial\beta}\right)\left(\frac{\partial L(y,u,\beta)}{\partial\beta}\right)^T\right] \qquad (3)$$

and $E_{y|\beta,u}$ denotes the conditional expectation over the distribution of y given β and u.

Notation

In the following, the arguments of L are suppressed where no confusion can arise.

If the estimator $\hat{\beta}$ is asymptotically efficient (e.g. maximum likelihood) then M_β^{-1} can be used as an approximation for cov $\hat{\beta}$ for long data lengths. The design problem then reduces to choosing a suitable scalar function of M_β^{-1} and selecting the sequence $\{u_k\}$ to minimise it.

Optimal experiment designs which are independent of the system and noise parameters exist for special classes of systems [L2] [F1]. In general, however, this does not occur and M_β depends on the true value of β which is unknown. To resolve this problem it is appropriate to adopt a Bayesian viewpoint and regard β as a random variable with a prior probability distribution. Although a good deal of controversy surrounds the use of Bayesian methods, the choice of prior distributions is often uncontroversial in control engineering applications [A8]. Payne and Goodwin [P1] have shown that the use of the prior mean $\bar{\beta}$ rather than the true value β is a good approximation provided the prior distribution is sufficiently sharp and $\bar{\beta}$ is sufficiently close to β. This is the approach adopted here.

2.4 FISHER INFORMATION MATRIX

For the purpose of constructing the log-likelihood function L, the noise sequence is assumed to be normally distributed and N observations of the input-output record are used to estimate β. It can then be shown that [G8]:

$$L = -\frac{N}{2} \log (d_0^2) - \frac{1}{2} \sum_{k=1}^{N} \varepsilon_k^2 + \text{constant} \tag{1}$$

where $\varepsilon_1, \ldots, \varepsilon_N$ is the residual sequence defined by

$$\varepsilon_k = \frac{C(z^{-1})}{D(z^{-1})} \{y_k - z^{-s} \frac{B(z^{-1})}{A(z^{-1})} u_k\} \tag{2}$$

2.4

From (1)

$$\frac{\partial L}{\partial \beta_i} = - \sum_{k=1}^{N} \varepsilon_k \frac{\partial \varepsilon_k}{\partial \beta_i} - \frac{N}{d_0} \delta_{i,p'-r} \qquad i = 1, \ldots, p' \qquad (3)$$

From (2)

$$\frac{\partial \varepsilon_k}{\partial a_i} = \frac{C(z^{-1})}{D(z^{-1})} \frac{B(z^{-1})}{A^2(z^{-1})} z^{-(s+i)} u_k, \qquad i = 1, \ldots, n \qquad (4)$$

$$\frac{\partial \varepsilon_k}{\partial b_i} = - \frac{C(z^{-1})}{D(z^{-1})} \frac{1}{A(z^{-1})} z^{-(s+i)} u_k, \qquad i = 0, \ldots, m$$

$$\frac{\partial \varepsilon_k}{\partial c_i} = \frac{z^{-i}}{C(z^{-1})} \varepsilon_k, \qquad i = 1, \ldots, q \qquad (5)$$

$$\frac{\partial \varepsilon_k}{\partial d_i} = - \frac{z^{-i}}{D(z^{-1})} \varepsilon_k, \qquad i = 0, \ldots, r$$

Note that:

(i) $\left\{ \frac{\partial \varepsilon_k}{\partial a_i} \right\}$ and $\left\{ \frac{\partial \varepsilon_k}{\partial b_i} \right\}$

do not depend on $\{\varepsilon_k\}$;

(ii) $\frac{\partial \varepsilon_k}{\partial c_i}$, $\frac{\partial \varepsilon_k}{\partial d_i}$

are statistically independent of ε_k for $i \geq 1$ and for all k;

(iii) $\left\{ \frac{\partial \varepsilon_k}{\partial c_i} \right\}$, $\left\{ \frac{\partial \varepsilon_k}{\partial d_i} \right\}$

are independent of $\{u_k\}$.

These considerations lead to the following simplified form of the Fisher information matrix, first derived by Payne [P5]:

$$M_\beta = E_{y|\beta,u} \left[\left(\frac{\partial L}{\partial \beta} \right) \left(\frac{\partial L}{\partial \beta} \right)^T \right]$$

$$= \begin{bmatrix} M & 0 \\ 0 & R \end{bmatrix} \tag{6}$$

where the partition of M_β corresponds to a partition of β between the system parameter vector Θ and the noise parameter vector Θ', i.e.

$$\begin{array}{cc} \beta^T & = & (\Theta^T \; \vdots \; \Theta'^T) \\ (1xp') & & (1xp) \quad (1xp'') \end{array}$$

$$\Theta^T = (a_1, \ldots, a_n, b_0, \ldots, b_m)$$

$$\Theta'^T = (c_1, \ldots, c_q, d_0, \ldots, d_r) \tag{7}$$

$$p = m + n + 1$$

$$p'' = q + r + 1$$

The pxp submatrix M is given by

$$M = \sum_{k=1}^{N} \left(\frac{\partial \varepsilon_k}{\partial \Theta} \right) \left(\frac{\partial \varepsilon_k}{\partial \Theta} \right)^T \tag{8}$$

where $\partial \varepsilon_k / \partial \Theta$ is given by (4). Clearly, each element of M is a quadratic function of the input sequence. However, the p''xp'' submatrix R is independent of $\{u_k\}$.

For large N,,

$$\text{cov } \hat{\beta} \simeq M_\beta^{-1} = \begin{bmatrix} M^{-1} & 0 \\ 0 & R^{-1} \end{bmatrix} \qquad (9)$$

so that

$$\text{cov } \hat{\theta} \simeq M^{-1}$$

$$\qquad (10)$$

$$\text{cov } \hat{\theta}' \simeq R^{-1}$$

Equation (10) shows that the accuracy with which the noise parameters can be estimated is not influenced by the choice of input sequence.

For all commonly used optimality criteria for input design (see Section 9), the noise submatrix in (9) contributes only an addition or multiplication constant to the cost function and hence has no effect on the optimization procedure. Therefore, in the following, only the system information matrix M will be considered in detail.

2.5 TIME DOMAIN DESIGN

In the time domain, the input design problem can be formulated as follows:

Find $u^* \in U$ s.t.

$$\Phi[M(u^*)] \leq \Phi[M(u)] \quad \forall \, u \in U \qquad (1)$$

where Φ is a scalar function expressing the chosen optimality criterion. Any $u^* \in U$ satisfying (1) is said to be Φ-optimal.

22

The set U is a subset of vectors in R^N, e.g.

$$U = \{u \,|\, u^T u \leq 1\} \tag{2}$$

corresponding to bounded input energy or

$$U = \{u \,|\, \alpha_k \leq u_k \leq \beta_k, \ k=1,\ldots,N\} \tag{3}$$

corresponding to (finite) input amplitude constraints. In both cases U is compact.

The need to search over a subset of R^N, where N is large, (e.g. Goodwin and Payne [G4]) is a computationally disadvantageous aspect of time-domain design theory. However, certain simplifications occur in the limit as $N \to \infty$. This limiting case is discussed in the next section.

2.6   FREQUENCY DOMAIN APPROACH

Assume that the limits

$$\bar{u} = \lim_{N \to \infty} \frac{1}{N} \sum_{k=1}^{N} u_k \tag{1}$$

and

$$r(\tau) = \lim_{N \to \infty} \frac{1}{N} \sum_{k=1}^{N} u_k u_{k+\tau}, \qquad \tau = 0, \pm 1, \ldots \tag{2}$$

exist and are finite.

Then, following Ljung [L4], the equations

$$r(\tau) = \int_{-\pi}^{\pi} e^{j\tau\omega} d\xi'(\omega), \qquad \tau = 0, \pm 1, \ldots \tag{3}$$

have a unique solution $\xi'(\omega)$ which is non-decreasing, right

continuous and has a derivative almost everywhere. The function

$\xi'(\omega)$ is the cumulative power distribution function of the input,

i.e. $\xi'(d\omega)$ is the input power in the frequency range $(\omega,\omega+d\omega]$ where

$\omega \in (-\pi,\pi]$.

Remark 1

If $\{u_k\}$ is a realization of a second order, ergodic stochastic

process, then $\bar{u}$ and $\{r(\tau)-\bar{u}^2, \tau=0,\pm1,...\}$ can be identified respectively

with the mean value and autocovariance of the process. Then ξ' is the

spectral distribution function.

It is more useful to work with the single-sided power distribution

function $\xi(\omega)$, defined on $[0,\pi]$ as follows [H1, p. 9]:

$$
\begin{aligned}
\xi(d\omega) &= 2\xi'(d\omega) \quad &\omega \in (0,\pi) \\
&= \xi'(d\omega) \quad &\omega = 0, \pi
\end{aligned}
\tag{4}
$$

In the context of experiment design, ξ will be referred to as the

design measure, the choice of which determines the relevant characteristics

of the input signal.

In general $\xi(\omega)$ will have a mixed spectrum, i.e. it will be

decomposable into components with continuous and discrete spectra

respectively.

Consider a general input with a discrete spectrum, i.e.

$$
u_k = \sum_{i=1}^{\ell} \alpha_i \sin (k\omega_i+\phi_i)
\tag{5}
$$

where

$$
0 = \omega_1 < \omega_2 < ... < \omega_\ell = \pi
$$

Then

$$\bar{u} = \alpha_1 \sin \phi_1 + \alpha_\ell \sin \phi_\ell \qquad (6)$$

$$r(\tau) = \sum_{i=1}^{\ell} \lambda_i \cos \tau \omega_i \qquad (7)$$

$$\xi(d\omega) = \sum_{i=1}^{\ell} \lambda_i \delta(\omega - \omega_i) d\omega \qquad (8)$$

where $\delta(\cdot)$ is the Dirac delta function and $\lambda_1, \ldots, \lambda_\ell$ are the input power proportions given by

$$\lambda_i = \alpha_i^2 \sin^2 \phi_i, \quad i = 1, \ell$$

$$= \tfrac{1}{2}\alpha_i^2 \quad \text{otherwise} \qquad (9)$$

Remark 2

If $\{u_k\}$ is a realization of discrete white noise, i.e.

$$\bar{u} = 0 ; \quad r(\tau) = 0, \ \tau \neq 0 \qquad (10)$$

then

$$\xi(d\omega) = \text{constant} \times d\omega = 2\xi'(d\omega) \qquad (11) \quad \#$$

In general, information will grow without bound as N increases and it can be shown [P2] that the elements of the information submatrix M increase linearly with N for long data lengths. It is therefore reasonable to consider the pxp asymptotic per sample information submatrix $\bar{M}$ defined by

$$\bar{M}_{ij} = \lim_{N \to \infty} \frac{1}{N} M_{ij}, \quad i, j = 1, \ldots, p \qquad (12)$$

2.6

where M is given by (4.4) and (4.8). (For brevity the matrix $\bar{M}$ will be referred to as 'information matrix' where no confusion is possible.)

In order to calculate $\bar{M}$, the equations (4.4) are written in the vector form:

$$\frac{\partial \varepsilon_k}{\partial \Theta} = h(z^{-1}) u_k \tag{13}$$

where

$$h_i(z^{-1}) = \frac{CB}{DA^2}(z^{-1}) z^{-(s+i)} \qquad i = 1, \ldots, n$$

$$= -\frac{C}{DA}(z^{-1}) z^{-(s+i-n-1)} \qquad i = n+1, \ldots, p \tag{14}$$

Notation

In (14) and the following, $A(z)B(z)$ is written $AB(z)$ and so on.

Application of Parseval's theorem [A10] then yields

$$\bar{M} = \lim_{N \to \infty} \frac{1}{N} \sum_{k=1}^{N} (\frac{\partial \varepsilon_k}{\partial \Theta})(\frac{\partial \varepsilon_k}{\partial \Theta})^T$$

$$= \int_{-\pi}^{\pi} h(e^{j\omega}) h^*(e^{j\omega}) d\xi'(\omega)$$

$$= \text{Re} \int_{0}^{\pi} h(e^{j\omega}) h^*(e^{j\omega}) d\xi(\omega) \tag{15}$$

where Re denotes 'real part of' and the superscript * denotes complex conjugate transpose.

Remark 3

Note that $\bar{M}$ does not depend on the time delay s.

For white noise input:

$$\bar{M} = \tfrac{1}{2} \alpha \int_{-\pi}^{\pi} h(e^{j\omega}) h*(e^{j\omega}) d\omega$$

$$= \frac{\alpha}{2j} \oint_{\substack{\text{unit} \\ \text{circle}}} h(z) h^{T}(z^{-1}) z^{-1} dz$$

$$= \pi\alpha \ \text{res} \ [h(z) h^{T}(z^{-1}) z^{-1}] \tag{16}$$

where α is the constant of proportionality in (11) and res $[\cdot]$ denotes the integral $(2\pi j)^{-1} \oint_{\substack{\text{unit} \\ \text{circle}}} [\cdot] dz$, i.e. the sum of the residues of $[\cdot]$ at its poles within the unit circle.

If the input spectrum is discrete, as in (5), then

$$\bar{M} = \text{Re} \ \sum_{i=1}^{\ell} \lambda_i h(e^{j\omega_i}) h*(e^{j\omega_i}) \tag{17}$$

2.7   POWER CONSTRAINTS

As in the time domain approach (Section 5) it is necessary to specify suitable constraints on the input sequence.

Definition 1

Denote by Ξ the set of ξ corresponding to the allowable set of inputs.

Here it is assumed that either the input power or output power is held constant. From (6.3) constant input power, P_0, requires that

$$\int_{-\pi}^{\pi} d\xi'(\omega) = P_0 = \int_{0}^{\pi} d\xi(\omega) \tag{1}$$

Definition 2

Denote by Ξ_1 the set of ξ for which

$$\int_0^\pi d\xi(\omega) = 1 \tag{2}$$

This corresponds to normalised input power.

Definition 3

Denote by $\mathcal{D}_1$ the subset of Ξ for which ξ has a purely discrete spectrum.

The matrix $\bar{M}$ is a linear functional of ξ and it is clear that the set of distributions for which (1) holds can be obtained from Ξ by scaling.

If $\xi \in \mathcal{D}_1$, as in (6.5), then (2) implies that

$$\sum_{i=1}^{\ell} \lambda_i = 1 \tag{3}$$

Input constraints may lead to outputs of unacceptable magnitude and it may be desirable to impose constraints on the output sequence.

For constant output power, P_0,

$$\lim_{N\to\infty} \frac{1}{N} \sum_{k=1}^{N} y_k^2 = P_0 \tag{4}$$

where $\{y_k\}$ is given by (2.6). Again, using Parseval's theorem, this yields

$$\int_0^\pi \frac{B(e^{j\omega})}{A(e^{j\omega})} \frac{B(e^{-j\omega})}{A(e^{-j\omega})} d\xi(\omega) + P_e = P_0 \tag{5}$$

where P_e is a constant given by

$$P_e = \text{res} \quad \frac{1}{z} \frac{D(z)}{C(z)} \frac{D(z^{-1})}{C(z^{-1})} \tag{6}$$

<u>Remark 1</u>

If $\{u_k\}$ is a realization of a second order, ergodic stochastic process, then (5) still holds provided $\{u_k\}$ and $\{e_k\}$ are uncorrelated sequences.

<u>Definition 4</u>

Denote by $\tilde{\Xi}$ the set of ξ for which $P_0 - P_e = 1$.

Writing the integrand in (5) as $d\eta(\omega)$, definition (4) implies that

$$\int_0^{\pi} d\eta(\omega) = 1 \tag{7}$$

It is clear that $\eta(\omega)$ is a valid design measure and that $\eta \in \Xi_1$.

From (6.13), $\bar{M}$ can be written as

$$\bar{M} = \text{Re} \int_0^{\pi} \tilde{h}(e^{j\omega}) \tilde{h}^*(e^{j\omega}) d\eta(\omega) \tag{8}$$

where

$$\tilde{h}(z) = A(z)h(z)/B(z) \tag{9}$$

These considerations show that the two types of constraint lead to identical design problems. This will be exploited later (Chapter 3, Section 8).

<u>Definition 5</u>

Denote by $\tilde{D}$ the subset of $\tilde{\Xi}$ for which ξ has a purely discrete spectrum.

It is clear that if $\xi \in \tilde{D}$ then $\eta \in D_1$. Then (7) takes the form of equation (3) where $\{\lambda_i\}$ is now the sequence of output power proportions. #

Design measures having a discrete spectrum are of particular importance (see next chapter) and are simple to realize. Consider the discrete design ξ^0 characterised by the power proportions $\lambda_1, \ldots, \lambda_\ell$ and frequencies $\omega_1, \ldots, \omega_\ell$ and corresponding to the input (6.5). Then, from (6.9),

$$u_k = \sqrt{\lambda_1} + \sum_{i=2}^{\ell-1} \sqrt{2\lambda_i} \sin(\omega_i k + \phi_i) + (-1)^k \sqrt{\lambda_\ell} \tag{10}$$

where the phase angles $\phi_2, \ldots, \phi_{\ell-1}$ are arbitrary.

If $\xi^0 \in \Xi_1$, then $P_0 \xi^0$ corresponds to a design with input power P_0 and is realized by the input sequence $\{P_0^{\frac{1}{2}} u_k\}$.

Let the design $(\lambda_1, \ldots, \lambda_\ell; 0, \omega_2, \ldots, \omega_{\ell-1}, \pi)$ now characterise the output design measure $\eta^0 \in \Xi_1$. Then the input sequence $\{(P_0 - P_e)^{\frac{1}{2}} u_k'\}$ yields total output power P_0 and corresponds to the design $(P_0 - P_e)^{\frac{1}{2}} \eta^0$, where

$$u_k' = \left|\frac{A(1)}{B(1)}\right| \sqrt{\lambda_1} + \sum_{i=2}^{\ell-1} \left|\frac{A(e^{j\omega_i})}{B(e^{j\omega_i})}\right| \sqrt{2\lambda_i} \sin(\omega_i k + \phi_i) + (-1)^k \left|\frac{A(-1)}{B(-1)}\right| \sqrt{\lambda_\ell} \tag{11}$$

and P_e is given by (6).

It is not necessary to use sinusoidal inputs to achieve a given input spectrum. Van den Bos [V4] shows how a desired spectrum can be approximated by a periodic binary signal. However, there exists a set of design measures that are not realizable by binary signals. Papoulis' arc sine law [P6, p. 483] yields exact binary realizations when this is possible.

Definition 6

Let M denote the set of information matrices corresponding to the set Ξ_1 of design measures.

The detailed structure of M is considered in the next chapter.

2.8   PERSISTENT EXCITATION

The minimal properties of test signals necessary for parameter identifiability have been rigorously analysed by a number of authors [A4][L4][S3][T1]. Roughly, the input must be sufficiently rich to excite all process modes of interest and such a signal is termed 'persistently exciting'. Ljung [L4] has shown that this implies a minimum number of spectral lines in the design measure while Rothenberg [R4] has shown that local parameter identifiability is equivalent to non-singularity of the information matrix.

The following theorem is of importance [M3].

Theorem 1

The information matrix $\bar{M}$ has the following properties:

(i) $\bar{M}$ is a real, symmetric nonnegative definite matrix;

(ii) $\bar{M}$ is singular if the input spectrum has less than $[p/2]$ lines where $[x]$ denotes the integer part of x.

Proof

(i) The result follows directly from the expression (6.15) for $\bar{M}$;

(ii) For an ℓ-frequency design $\{\lambda_i, \omega_i, \ i=1,\ldots,\ell\}$

$$\bar{M} = \text{Re} \sum_{i=1}^{\ell} \lambda_i h(e^{j\omega_i}) h^*(e^{j\omega_i}) \tag{1}$$

The real part of an outer product of two vectors has at most rank 2. It follows from (1) that

$$\text{rank } \bar{M} \leq 2\ell \tag{2}$$

However the matrix $\bar{M}$ is singular iff rank $\bar{M} < p$ and the result follows. #

Part (ii) of the theorem gives a necessary condition for local identifiability, i.e. that the input spectrum must exhibit at least $[p/2]$ lines. It is now shown that the specific structure of $\bar{M}$ can be exploited to derive a necessary and sufficient condition for local identifiability.

Definition 1

If ξ has a purely discrete spectrum, define a *design index* $I(\xi)$ as the number of design frequencies in $[0,\pi]$ where the end frequencies 0 and π are each counted as one half.

* Theorem 2

The information matrix $\bar{M}(\xi)$ is nonsingular iff the design index $I(\xi) \geq p/2$.

Proof

Necessity. This requires a slight modification of the proof of Theorem 1, part (ii).

Note that rank $[h(e^{j\omega})h*(e^{j\omega})] = 1$ if $\omega = 0$ or π. Consider ξ to be the ℓ-frequency design leading to (1) and assume, without loss of generality, that

$$\lambda_i > 0, \quad i = 1, 2, \ldots, \ell$$

$$0 \leq \omega_1 < \omega_2 < \ldots < \omega_\ell \leq \pi$$

Inequality (2) is replaced by

$$\text{rank } \bar{M}(\xi) \le 2\ell \text{ if } \omega_1 \ne 0, \ \omega_\ell \ne \pi, \text{ i.e. } I(\xi) = \ell$$

$$\le 1 + 2(\ell-1) \text{ if } \omega_1 = 0 \text{ or } \omega_\ell = \pi, \text{ i.e. } I(\xi) = \ell - \tfrac{1}{2}$$

$$\le 2 + 2(\ell-2) \text{ if } \omega_1 = 0, \ \omega_\ell = \pi, \text{ i.e. } I(\xi) = \ell - 1$$

i.e., in general,

$$\text{rank } \bar{M}(\xi) \le 2I(\xi)$$

Therefore, if $I(\xi) < p/2$, $\bar{M}(\xi)$ is singular. Hence necessity is proved.

Sufficiency. Assume that ξ is an ℓ-frequency design as before and that $I(\xi) \ge p/2$. Assume also that $\bar{M}(\xi)$ is singular. Then the equation $\alpha^T \bar{M}\alpha = 0$, where $\alpha \underline{\Delta} (\alpha_1, \ldots, \alpha_p)^T$ is a real p-vector, has a nontrivial solution.

From (1)

$$\alpha^T \bar{M}\alpha = \sum_{i=1}^{\ell} \lambda_i |\alpha^T h(e^{j\omega_i})|^2 \tag{3}$$

therefore

$$\alpha^T \bar{M}\alpha = 0 \Rightarrow \alpha^T h(e^{j\omega_i}) = 0, \quad i = 1, \ldots, \ell \text{ if } \alpha \ne 0$$

and

$$\alpha^T h(e^{j\omega_i}) = 0 \Rightarrow \alpha^T h(e^{-j\omega_i}) = 0, \quad i = 1, \ldots, \ell$$

Therefore $\alpha^T h(z) = 0$ at ℓ' distinct point on the unit circle where

$$\ell' = 2\ell \text{ if } \omega_1 \neq 0, \ \omega_\ell \neq \pi$$

$$= 2\ell - 1 \text{ if } \omega_1 = 0 \text{ or } \omega_\ell = \pi$$

$$= 2\ell - 2 \text{ if } \omega_1 = 0, \ \omega_\ell = \pi$$

From Definition (1)

$$\ell' = 2I(\xi) \tag{4}$$

From (6.14)

$$\alpha^T h(z) = \frac{C}{DA^2}(z) z^s \{B(z) \sum_{i=1}^{n} \alpha_i z^i - A(z) \sum_{i=0}^{m} \alpha_{i+n+1} z^i\} \tag{5}$$

The right hand side of (5) has at most $p - 1$ roots on the unit circle. Therefore, from (4), if $I(\xi) \geq p/2$ then $\alpha^T h(z) \equiv 0$. This implies that either

(a) $\quad A(z) \equiv \sum_{i=1}^{n} \alpha_i z^i$

$\quad B(z) \equiv \sum_{i=0}^{m} \alpha_{i+n+1} z^i$

apart from possible constant factors, or

(b) $\quad \sum_{i=1}^{n} \alpha_i z^i \equiv 0 \equiv \sum_{i=0}^{m} \alpha_{i+n+1} z^i$

i.e.

$$\alpha = 0$$

Case (a) cannot occur because $A(z)$ has a constant term equal to unity.

Therefore $\alpha^T \bar{M} \alpha = 0 \Rightarrow \alpha = 0$ if $I(\xi) \geq p/2$ and this contradicts the assumed singularity of $\bar{M}(\xi)$. This implies that if $I(\xi) \geq p/2$, then $\bar{M}(\xi)$ is nonsingular.   q.e.d.#

Corollary

If $I(\xi) \leq p/2$, rank $\bar{M}(\xi) = 2I(\xi)$.

The last step in setting up the input design problem is to choose a suitable measure of design efficiency. This is considered in the next section.

2.9   CHOICE OF COST FUNCTION

A measure of efficiency in an identification experiment can be expressed as a scalar function $\Phi[\bar{M}]$.

Definition 1

A design measure $\xi^* \in \Xi$ is said to be Φ-optimal if

$$\Phi[\bar{M}(\xi^*)] \leq \Phi[\bar{M}(\xi)] \quad \forall \ \xi \in \Xi \tag{1}$$

Remark 1

Clearly the direction of the inequality in (1) is convention and can always be reversed by redefining Φ.

The particular form of Φ should reflect the use to which the process model will be put and this leads to a variety of choices [F1] [M2]. A desirable property is that of ordering the designs, i.e.

$$\Phi[\bar{M}(\xi_1)] < \Phi[\bar{M}(\xi_2)] \text{ if } \bar{M}(\xi_1) > \bar{M}(\xi_2) \tag{2}$$

(i.e. $\bar{M}(\xi_1) - \bar{M}(\xi_2)$ is positive definite).

A number of authors [A2][A3][N2][N3][N4][M1] use cost functions which are variants of

$$\Phi[\bar{M}] = \{\text{trace } (W\bar{M})\}^{-1} \qquad (3)$$

where W is a symmetric, nonnegative definite weighting matrix. This criterion leads to a straightforward quadratic optimization problem but it has been pointed out [Z1][G3][R3][T1][M2] that the resulting optimal input may not be persistently exciting. It is now shown that this criticism holds for a wider class of optimality criteria.

Definition 2

A design $\xi^* \in \Xi$ is said to be L-optimal if (1) holds and Φ satisfies the condition

$$\{\Phi[\alpha_1\bar{M}_1+\alpha_2\bar{M}_2]\}^{-1} = \alpha_1\{\Phi[\bar{M}_1]\}^{-1} + \alpha_2\{\Phi[\bar{M}_2]\}^{-1} \qquad (4)$$

for arbitrary scalars α_1, α_2.

Clearly, cost function (3) belongs to the class of L-optimality criteria.

* Theorem 1

If $\xi \in \Xi_1 \cup \tilde{\Xi}$, then L-optimality is achieved by a single frequency design.

Proof

If $\xi \in \Xi \cup \tilde{\Xi}$, then

$$\int_0^\pi \alpha(\omega)\,d\xi(\omega) = 1 \qquad (5)$$

where

$$\alpha(\omega) = 1 \text{ if } \xi \in \Xi_1$$

$$= |B(e^{j\omega})/A(e^{j\omega})|^2 \text{ if } \xi \in \tilde{\tilde{\Xi}}$$

and α is therefore continuous, finite and positive in $[0,\pi]$.

The matrix $\bar{M}$ is given by

$$\bar{M} = \text{Re} \int_0^{\pi} h(e^{j\omega}) h*(e^{j\omega}) d\xi(\omega) \tag{6}$$

Introducing the measure

$$d\eta(\omega) = \alpha(\omega) d\xi(\omega) \tag{7}$$

then, for any Φ satisfying (4),

$$\{\Phi(\bar{M})\}^{-1} = \int_0^{\pi} \phi(\omega) d\eta(\omega) \tag{8}$$

where

$$\int_0^{\pi} d\eta(\omega) = 1 \tag{9}$$

and

$$\phi(\omega) = \{\Phi[\text{Re } h(e^{j\omega}) h*(e^{j\omega})]\}^{-1} \tag{10}$$

From (8) and (9) it is clear that L-optimality is achieved by $\eta*$ given by

$$d\eta*(\omega) = \delta(\omega-\omega*) d\omega \tag{11}$$

where

$$\omega* = \arg \inf_{\omega \in [0,\pi]} \phi(\omega) \qquad \text{Q.E.D. \#}$$

Remark 2

L-optimality may be achieved by other designs. If ξ_1, ξ_2 are both L-optimal and the input constraint is of the form (5) then any convex combination of ξ_1, ξ_2 is L-optimal.

The design index $I(\xi)$ for a single frequency design ξ is either 1/2 or 1. It follows from Theorem (8.2) that

$$\bar{M}(\xi) \text{ is singular if } p > 1 \text{ for } I(\xi) = 1/2$$

$$\text{if } p > 2 \text{ for } I(\xi) = 1$$

It is clear that, if identifiability problems are to be avoided, the L-optimality criterion should not be used for $p \geq 3$. This class of criteria are excluded by demanding that

$$\lim_{\det \bar{M} \to 0} \Phi[\bar{M}] = \infty \tag{12}$$

This condition is satisfied by the most common choices for Φ:

(a) log det $\bar{M}^{-1}$: D-optimality. An equivalent criterion is det $\bar{M}^{-1}$.

Payne [P2] has shown that this criterion can be derived from Lindley's average information increment [L5] in the case of normal distributions and can be used as a measure of the return from an experiment when the goal is solely to gain information (rather than, say, to design a controller). A related concept is that of mutual information between parameters and data used by Arimoto and Kimura [A9].

In Section (4.4) it is shown that D-optimality is equivalent to a certain output space criterion. In particular, if the process noise is white, a D-optimal input minimises the variance of the one-step-ahead prediction error.

An important property of D-optimal designs is their invariance with respect to parameter transformations with nonsingular Jacobians [M3].

(b)   Trace $\overline{W}\overline{M}^{-1}$: L-optimality. The matrix W is a symmetric nonnegative

definite weighting matrix. If W is diagonal the cost function is a

weighted sum of the parameter variances. If W is the unit matrix, the

criterion is usually called A-optimality.

Payne [P2] shows that this criterion can be derived as an

approximation to more general cost criteria when third and higher order

moments of the prior distribution function of the parameters can be

neglected.

(c) $\lambda_{max}(\overline{M}^{-1})$, i.e. the maximum eigenvalue of $\overline{M}^{-1}$: E-optimality.

A more general cost function has been constructed by Müller and

Weber [M5]:

$$\Phi_s(\overline{M}) = [\tfrac{1}{p} \text{ trace } \overline{M}^{-s}]^{1/s}, \ s = 0, \ 1, \ ... \tag{13}$$

It can be shown that Φ_0, Φ_1, Φ_∞ are equivalent to the D-, A- and

E-optimality criteria respectively. Also

$$\Phi_{s_1} \geq \Phi_{s_2} \text{ if } s_1 \geq s_2 \tag{14}$$

This inequality implies that, for a given design, λ_{max} is an upper

bound on all other criteria in this class, i.e. if λ_{max} is 'small', no

Φ_s can be 'large'. This suggests that E-optimality is preferable to

other choices, however this must be weighted against the computational

disadvantages of eigenvalue calculations.   #

Definition 3

Φ is *convex* if

$$\Phi[\alpha\bar{M}_1 + (1-\alpha)\bar{M}_2] \leq \alpha\Phi[\bar{M}_1] + (1-\alpha)\Phi[\bar{M}_2], \; \forall \; \alpha \; \epsilon \; (0,1) \qquad (15)$$

and *strictly convex* if, in addition, $\bar{M}_1 = \bar{M}_2$ is a necessary condition

for equality in (15).

It is not difficult to show that Φ_s is strictly convex. In

general, trace $\overline{WM}^{-1}$ is convex but only strictly convex if W is

nonsingular.

It is assumed that any cost function of interest is (at least)

convex, satisfies conditions (2) and (12) and is continuously

differentiable in the neighbourhood of $\bar{M}(\xi^0)$, where ξ^0 is any design

of interest (in particular, the Φ-optimal design).

More general cost criteria are discussed by Fedorov [F1] and

Kiefer [K6].

This completes the formulation of the input design problem. In

the next chapter the set M of information matrices corresponding to

input power constraint is analysed in detail from a geometrical

standpoint.

2.10 CONCLUDING REMARKS

In this chapter, the problem of optimal input design for a

linear SISO dynamic system has been formulated from a frequency domain

standpoint. Theoretically, the steady state approach demands an

infinite data length. Gustavsson [G2] suggests that preliminary

experiments should have a duration of at least ten times the longest

time constant. In practical aircraft parameter identification

greater than two time cycles for the short period mode of the
aircraft longitudinal motions. This corresponds to 500 data points.
For data lengths shorter than this, the phase relationships between
different frequency components become important.

Chapter 3

A TCHEBYCHEFF SYSTEM APPROACH

3.1   INTRODUCTION

A geometrical approach to the analysis of M, the set of information matrices corresponding to normalised input power, is developed in Sections 2 and 3. Payne [P3] has shown that M is isomorphic to a set $M^{(p)}$ in R^p and this leads to an upper bound p on the number of input frequencies necessary to attain a Φ-optimum. Conditions are sought under which there exists a Φ-optimal design ξ^* for which $I(\xi^*) < p$ and in particular $I(\xi^*) = p/2$, the minimum necessary to ensure identifiability.

The theory of Tchebycheff systems furnishes a fruitful approach to these problems (Sections 4 and 5) whereby $M^{(p)}$ is imbedded in a moment space $M_c^{(p)}$, a closed convex cone induced in R^p by a T-system $\{v_i\}_1^p$ on $[0,\pi]$. The T-system arises naturally out of Payne's theorem. These considerations lead to the concept of canonical and principal representations with the required minimal properties (Section 6) and to sufficient conditions for the existence of optimal designs of the desired type, placing only weak restrictions on the optimality criteria (Sections 6 and 7).

Theorems (5.2) and (6.4) establish that $M^{(p)}$ is a hyperplane in R^p iff q = 0, m $\geq$ n+r and that this is sufficient for the existence of a Φ-optimal design ξ^*, unique on arbitrarily fixing one spectral frequency and for which $I(\xi^*) \leq (p+1)/2$. Further, Theorem (7.1) establishes an upper bound $\frac{1}{2}$ max (p+q-1, 2n+r) on $I(\xi^*)$ if $M^{(p)}$ is not a hyperplane and and this leads directly to sufficient conditions for optimal designs to exist for which $I(\xi^*) \leq (p+1)/2$ [Theorem (7.2)].

The case of normalised output power is considered in Section 8 and it is shown that the analysis is based on a different T-system but otherwise parallels that of the previous sections. The hyperplane condition is found to be q = 0 = r [Theorem (8.1)] and the upper bound on the optimal design index is $\frac{1}{2}$(p+1) + $\frac{1}{2}$ max (r,q) [Theorem (8.3)].

Appendix A presents some examples to illustrate the main body of theory presented in this chapter. Javaherian [J1] has conjectured that optimal designs may always be constructed from [(p+1)/2] frequencies. In Appendix B the geometrical approach is used to construct counter-examples to this conjecture.

3.2   SOME PROPERTIES OF M

Consider $\bar{M}$ ϵ M.   Then

$$\bar{M} = \text{Re} \int_0^\pi h(e^{j\omega}) h^*(e^{j\omega}) d\xi(\omega) \qquad (1)$$

where ξ ϵ Ξ_1, the set of design measures corresponding to the normalised input power, i.e.

$$\int_0^\pi d\xi(\omega) = 1 \qquad (2)$$

The following theorem sets out some important properties of
the set M and emphasises the crucial position occupied by the subset
D_1 of Ξ_1 corresponding to design measures satisfying (2) and with
discrete spectra.

Theorem 1 [K8][M3]

(i) The set M is compact and is the convex hull of the subset of
M corresponding to single frequency designs;

(ii) if $\xi_1 \epsilon \Xi_1$, then these exists $\xi_2 \epsilon D_1$ whose spectrum contains
no more than $[p(p+1)/2+1]$ points and such that $\bar{M}(\xi_1) = \bar{M}(\xi_2)$. If
$\bar{M}(\xi_1)$ lies on the boundary of M then no more than $p(p+1)/2$ points are
required.

Proof

(i) This follows closely the proof given by Karlin and Studden
for static experiment design [K8, p. 324].

Let $\ell = p(p+1)/2$ and denote by Ω the compact set $\Omega = \{\omega \mid 0 \leq \omega \leq \pi\}$.

Let $z_1(\omega), \ldots, z_\ell(\omega)$ denote the set of continuous functions
Re $\lceil h_1(e^{j\omega}) h_k(e^{-j\omega}) \rceil$ $(k=1,\ldots,i;$ $i=1,\ldots,p)$ arranged in some order
and introduce the set $M^{(\ell)} \subset R^\ell$ defined by

$$M^{(\ell)} = \{(x_1, \ldots, x_\ell) \mid x_i = \int_\Omega z_i(\omega) \, d\xi(\omega), \ i=1,\ldots,\ell, \xi \epsilon \Xi_1\} \quad (3)$$

It is clear that there exists a linear mapping between M and $M^{(\ell)}$
and that the two sets are isomorphic.

Consider the single frequency design curve, i.e. the set in R^ℓ
defined by

$$C_\ell = \{(z_1(\omega), \ldots, z_\ell(\omega)) \mid \omega \epsilon \Omega\} \quad (4)$$

and denote by $C(c_\ell)$ the convex hull of c_ℓ. Then Caratheodory's theorem [F1, p.66] states that each $\underline{x}$ in $C(c_\ell)$ possesses a representation of the form

$$x_i = \sum_{k=1}^{\ell+1} \alpha_k z_i(\omega_k), \quad i = 1, \ldots, \ell \tag{5}$$

where

$$\alpha_k \geq 0, \quad k = 1, \ldots, \ell+1$$

$$\sum_{k=1}^{\ell+1} \alpha_k = 1$$

and $\alpha_{\ell+1}$ can be set equal to zero if $\underline{x}$ lies on the boundary of $C(c_\ell)$.

Since Ω is compact and $z_1, \ldots, z_\ell$ are continuous, the representation (5) implies that $C(c_\ell)$ is compact. Further $C(c_\ell)$ closed implies $C(c_\ell) = M^{(\ell)}$ (see [K8, pp. 145-6] for proof). This completes the proof of (i).

(ii) This follows from (3), (5) and Caratheodory's theorem stated above.   #

Remark 1

Each matrix in M is represented as a point in a finite dimensional Euclidean space. This invites a geometric approach to the study of the structure of M through analysis of $M^{(\ell)}$ and this is exploited later.

Remark 2

Part (ii) of the theorem is particularly important for optimal input design. Any $\bar{M} \in M$ can be represented in the form

$$\bar{M} = \text{Re} \sum_{i=1}^{\ell+1} \lambda_i h(e^{j\omega_i}) h^*(e^{j\omega_i}) \tag{6}$$

where $\ell = p(p+1)/2$ and $\{\lambda_i, i=1,\ldots,\ell+1\}$ is a non-negative sequence satisfying

$$\sum_{i=1}^{\ell+1} \lambda_i = 1 \tag{7}$$

Therefore the search for a Φ-optimal design involves at most the $[p(p+1)+2]$ search variables $(\lambda_1,\ldots,\lambda_{\ell+1},\omega_1,\ldots,\omega_{\ell+1})$, in general far less than the number of data points (compare the time domain case). The optimal inputs can be realised by a combination of sine waves (see Section 2.7).   #

The theorem also leads to some useful properties of the Φ-optimal design $\xi*$.

I. The existence of $\xi*$ is guaranteed by the compactness of M and the continuity of Φ (see, for example, Luenberger [L6, pp. 39-40]).

II. If ξ^0 yields a local minimum for Φ, then $\Phi[\bar{M}(\xi^0)] = \Phi[\bar{M}(\xi*)]$. This follows from the convexity of Φ and the fact that $\bar{M}(\xi)$ is linear in ξ [L6, p. 191].

III. $\bar{M}(\xi*) \in \text{Bd } M$. Assume that $\bar{M}* \triangleq \bar{M}(\xi*) \in \text{Int } M$. Then $(1+\alpha)\bar{M}* \in \text{Int } M$ for some $\alpha > 0$. But $(1+\alpha)\bar{M}* > \bar{M}*$. Therefore $\Phi[(1+\alpha)\bar{M}*] < \Phi[\bar{M}*]$ from condition (2.9.2). This contradicts the Φ-optimality of $\xi*$.

IV. If Φ is strictly convex, then $\bar{M}*$ is unique. Suppose that $\Phi[\bar{M}*] = \Phi[\bar{M}_1]$ but $\bar{M}_1 \neq \bar{M}*$. Then, for any $\alpha \in (0,1)$, $\Phi[\alpha\bar{M}*+(1-\alpha)\bar{M}_1] < \alpha\Phi[\bar{M}*] + (1-\alpha)\Phi[\bar{M}_1] = \Phi[\bar{M}*]$, contradicting the Φ-optimality of $\xi*$.

Remark 3

Property III implies that the number of search variables in the design problem can be further reduced to at most $p(p+1)$.

Remark 4

If Φ is not strictly convex, then $\bar{M}*$ is not necessarily unique.

3.3   PAYNE'S THEOREM

The ability to represent any $\bar{M} \in M$ as a point in $R^{p(p+1)/2}$ is not dependent on the detailed structure of $\bar{M}$. Any pxp symmetric matrix can be represented in this way. However, exploiting the assumed model structure (2.1.6) and (2.1.7), $\bar{M}$ is given by:

$$\bar{M} = \text{Re} \int_0^\pi h(e^{j\omega}) h*(e^{j\omega}) d\xi(\omega) \tag{1}$$

where

$$h_i(z) = \frac{CB}{DA^2}(z) z^{s+i} \qquad i = 1, \ldots, n$$

$$= -\frac{C}{DA}(z) z^{s+i-n-1} \qquad i = n+1, \ldots, p \tag{2}$$

and it can be shown that [P3]:

Theorem 1 [Payne]

The matrix $\bar{M}$ can be represented as a point in R^p.

Proof

Let

$$f(\omega) = \frac{C}{DA^2}(e^{j\omega}) \frac{C}{DA^2}(e^{-j\omega}) \tag{3}$$

Then

$$h_i(e^{j\omega}) h_k(e^{-j\omega}) = f(\omega) B(e^{j\omega}) B(e^{-j\omega}) e^{j\omega(i-k)} \qquad i,k = 1,\ldots,n$$

$$= -f(\omega) B(e^{j\omega}) A(e^{-j\omega}) e^{j\omega(i-k+n+1)} \qquad \begin{array}{l} i = 1,\ldots,n \\ k = n+1,\ldots,p \end{array}$$

$$= -f(\omega) B(e^{-j\omega}) A(e^{j\omega}) e^{j\omega(i-k-n-1)} \qquad \begin{array}{l} i = n+1,\ldots,p \\ k = 1,\ldots,n \end{array}$$

$$= f(\omega) A(e^{j\omega}) A(e^{-j\omega}) e^{j\omega(i-k)} \qquad i, k = n+1,\ldots,p$$

(4)

Inspection of equations (4) and use of de Moivre's theorem leads to the expression

$$\bar{M} = \sum_{i=1}^{p} x_i L_i \qquad (5)$$

where $L_1, \ldots, L_p$ are constant pxp matrices depending only on the coefficients of the polynomials $A(z)$, $B(z)$ and $x_1, \ldots, x_p$ are scalars which depend on the design measure and are given by

$$x_i = \int_0^{\pi} f(\omega) \cos^{i-1} \omega \, d\xi(\omega) \qquad i = 1, \ldots, p \qquad (6)$$

This is the required result.   #

Remark 1

Payne proves the result for a continuous-time system, however the proof is exactly the same as above.

Theorem 1 implies a considerable reduction in the number of search variables to at most 2p for the input design problem, e.g. for p = 10 a reduction from 110 to 20. However, it is clear from theorem (2.8.1) that this is still approximately double the theoretical minimum necessary to ensure identifiability.

The question arises: Under what conditions does a Φ-optimal design ξ^* exist for which $I(\xi^*) < p$ and in particular $I(\xi^*) = p/2$? (The corresponding problem in static experimental design has been designated an 'open problem' by Karlin and Studden [K8, p. 373].)

The main theoretical results in the following sections are directed towards a partial answer to this question.

3.4   TCHEBYCHEFF SYSTEM APPROACH

Definition 1

Let v_1, v_2, ..., v_p denote continuous real-valued functions defined on a closed finite interval $[a,b]$. These functions are called a *Tchebycheff system over* $[a,b]$, (or *T-system*), provided the pth order determinants

$$v \begin{pmatrix} 1,2,\dots,p \\ t_1,t_2,\dots,t_p \end{pmatrix} \triangleq \begin{vmatrix} v_1(t_1) & v_1(t_2) & \cdots & v_1(t_p) \\ v_2(t_1) & v_2(t_2) & \cdots & v_2(t_p) \\ \vdots & & & \vdots \\ v_p(t_1) & v_p(t_2) & \cdots & v_p(t_p) \end{vmatrix} \quad (1)$$

are strictly positive whenever $a \le t_1 < t_2 < \dots < t_p \le b$. Also let $\underline{v}$ denote the vector $col(v_1,\dots,v_p)$ and $\{v_i\}_1^p$ denote the sequence of functions v_1, v_2, ..., v_{p-1}, $-v_p$.

Remark 1

The Tchebycheff property implies that the vectors $\underline{v}(t_1)$, ..., $\underline{v}(t_k)$ form a linearly independent set if $k \le p$ and $a \le t_1 < t_2 < \dots < t_k \le b$.

* Theorem 1

Let $v_i(\omega) = f(\omega) \cos^{i-1}\omega$, $i = 1$, ..., p, $\omega \in [0,\pi]$ where $f(\omega)$ is given by (3.3). Then $\{v_i\}_1^p$ is a T-system on $[0,\pi]$ if $p(p-1)/2$ is even. Otherwise $\{v_i^-\}_1^p$ is a T-system on $[0,\pi]$.

Proof

From (1)

$$V \begin{pmatrix} 1,2,\ldots,p \\ \omega_1,\omega_2,\ldots,\omega_p \end{pmatrix} = \begin{vmatrix} f(\omega_1) & \ldots & f(\omega_p) \\ f(\omega_1)\cos\omega_1 & \ldots & f(\omega_p)\cos\omega_p \\ \vdots & & \\ f(\omega_1)\cos^{p-1}\omega_1 & \ldots & f(\omega_p)\cos^{p-1}\omega_p \end{vmatrix}$$

$$= \left\{ \prod_{i=1}^{p} f(\omega_i) \right\} \prod_{i \le j < k \le p} (\cos\omega_k - \cos\omega_j) \quad (2)$$

where $0 \le \omega_1 < \omega_2 < \ldots < \omega_p \le \pi$.

The polynomials A, C, D do not vanish on the unit circle (by assumption). Therefore $\{v_i\}_1^p$ is a system of continuous functions and $f(\omega) > 0$, $\forall \omega \in [0,\pi]$. Each term in the second product of (2) is negative. Hence it follows that

$$V \begin{pmatrix} 1,2,\ldots,p \\ \omega_1,\omega_2,\ldots,\omega_p \end{pmatrix} > 0 \text{ if } p(p-1)/2 \text{ is even}$$
$$< 0 \text{ if } p(p-1)/2 \text{ is odd}$$

The use of $\{v_i^-\}_1^p$ changes the sign of every determinant. Therefore, if $p(p-1)/2$ is odd, $\{v_i^-\}_1^p$ is a T-system on $[0,\pi]$. q.e.d. #

Remark 2

Either $\{v_i\}_1^p$ or $\{v_i^-\}_1^p$ is a T-system on $[a,b]$ iff $0 \le a < b \le \pi$.

A further result of use (see Main Appendix) is:

* Result 1

A necessary and sufficient condition for both $\{v_i\}_1^{p-1}$ and $\{v_i\}_1^p$ to be T-systems on $[0,\pi]$ or $\{v_i^-\}_1^{p-1}$ and $\{v_i^-\}_1^p$ to be T-systems on $[0,\pi]$ is that p is odd.

Proof

If $p(p-1)/2$ and $(p-1)(p-2)/2$ are either both even or both odd then their difference, $p-1$, is even, i.e. p is odd. This establishes sufficiency. Similarly for necessity.   q.e.d. #

The Tchebycheff property established in Theorem 1 allows the use
of a well-established body of theory in the following. This theory
is fully developed in the book of Karlin and Studden [K8] and the
main results are brought together for ease of reference in the Main
Appendix to this thesis. Reference will be made to this appendix
where necessary. It may also be helpful to the reader to refer to
the examples in Appendix A as illustrations of the ideas presented
in the remainder of this chapter.

3.5   MOMENT SPACES

The set of information matrices M_Ξ corresponding to some
admissible set Ξ of design measures is isomorphic to the subset
$M_\Xi^{(p)}$ of R^p defined by

$$M_\Xi^{(p)} = \{\underline{x} \epsilon R^p \mid \underline{x} = \int_0^\pi \underline{v}(\omega)\, d\xi(\omega), \xi \epsilon \Xi\} \tag{1}$$

using (3.5) and (3.6).

Notation

Extending the notation introduced in (1), the superfix (p) is
used in the following to label a domain in R^p of the mapping (3.5),
i.e. $S^{(p)} \to S$ under the mapping where $S^{(p)} \subset R^p$ and $S \subset R^{p \times p}$, the space
of pxp matrices.

In the classical theory [K8, Chapter 2], $M_\Xi^{(p)}$ is a *moment space*
induced by the T-system $\{v_i\}_1^p$. If $\Xi = \Xi_1$, then $M_\Xi = M$, the set of
immediate interest. However it is more useful to imbed M in a larger
set M_c obtained by choosing Ξ as the set of all nondecreasing right

continuous functions of bounded variation.

By definition

$$M_c^{(p)} = \{ \lambda \underline{x} \mid \lambda \geq 0, \ \underline{x} \in M^{(p)} \}$$

i.e. $M_c^{(p)}$ is a cone in R^p.

Theorem 1 [K8, pp. 38-40]

The moment space $M_c^{(p)}$ is a closed convex cone.   #

The proof of Theorem 1 follows closely that of Theorem (2.1).

In fact there is a close relationship between $M_c^{(p)}$ and $M^{(p)}$.

Result 1

Every ray of $M_c^{(p)}$ passes through at least one point of $M^{(p)}$.

Proof

Let $\underline{x} \in M_c^{(p)}$ and correspond to a measure ξ for which $\int_0^\pi d\xi(\omega) = \alpha > 0$.

Then $\alpha^{-1}\xi$ is a measure for which $\int_0^\pi d[\alpha^{-1}\xi](\omega) = 1$, i.e. $\alpha^{-1}\underline{x} \in M^{(p)}$.   #

The sets $M^{(p)}$ and $M_c^{(p)}$ can therefore be generated in the following

way:

   (i)   Construct the trajectory C_p of single frequency designs,

i.e.

$$C_p = \{\underline{v}(\omega) \mid 0 \leq \omega \leq \pi\} \qquad\qquad (2)$$

   (ii)   Form the convex hull of C_p, i.e. $M^{(p)}$ (Theorem 2.1).

   (iii)   Then $M_c^{(p)}$ is the set of all rays passing through $M^{(p)}$.

In general, a ray of $M_c^{(p)}$ will pass through an infinite number of

points of $M^{(p)}$. However if $M^{(p)}$ is a hyperplane a ray will intersect

at a unique point. The condition for this to occur is given in Theorem 2.

Recall that the structure of the process model (2.2.6) is

determined by a time delay s and the polynomials A, B, C, D of orders

n, m, q, r respectively.

3.5

* Theorem 2

The moment space $M^{(p)}$ is a hyperplane in R^p iff the orders of the polynomials in the process model (2.2.6) satisfy the conditions:

$$q = 0, \quad m \geq n+r \qquad (3) \dagger$$

Proof

Let $\underline{x} \in M^{(p)}$. Then, for arbitrary $\underline{\alpha} = (\alpha_1, \ldots, \alpha_p)^T$, $\exists \xi \in \Xi_1$, s.t.

$$\sum_{i=1}^{p} \alpha_i x_i = \int_0^{\pi} \{ \sum_{i=1}^{p} \alpha_i v_i(\omega) \} d\xi(\omega) \qquad (4)$$

But

$$\sum_{i=1}^{p} \alpha_i v_i(\omega) = f(\omega) \sum_{i=1}^{p} \alpha_i \cos^{i-1}\omega$$

$$= \frac{pol_q(t)}{pol_{2n+r}(t)} \sum_{i=1}^{p} \alpha_i t^{i-1} \qquad (5)$$

where $t \underline{\Delta} \cos \omega$ and $pol_q(t)$ denotes a polynomial of degree q in t. By assumption A, B, C, D are relatively prime and therefore so are pol_q and pol_{2n+r}.

The moment space $M^{(p)}$ is a hyperplane

$\Leftrightarrow \exists \underline{\alpha}$ s.t. $\sum_{i=1}^{p} \alpha_i x_i = $ constant $\forall \underline{x} \in M^{(p)}$

$\Leftrightarrow \exists \underline{\alpha}$ s.t. the R.H.S. of (4) is constant $\forall \xi \in \Xi_1$

$\Leftrightarrow \exists \underline{\alpha}$ s.t. the R.H.S. of (5) is constant $\forall t \in [-1,1]$

$\Leftrightarrow \exists \underline{\alpha}$ s.t. the numerator and denominator polynomials in (5) have the same zeros

$\Leftrightarrow q = 0, p-1 \geq 2n+r$ where $p = m+n+1$

$\Leftrightarrow$ condition (3) is satisfied. q.e.d. #

$\dagger$A commonly used canonical form for the process model is
$$A(z^{-1})y_k = B(z^{-1})u_k + C(z^{-1})e_k$$
where the polynomials A,B,C are all of degree n. For this case, a suitable moment space can be constructed and it can be shown that the relevant hyperplane conditions are *always* satisfied for both input and output power constraints.

An important case in which (3) is satisfied occurs when the output is a moving average of the input corrupted by white noise, i.e. only m is non-zero so that

$$y_k = z^{-s}B(z^{-1})u_k + e_k \qquad\qquad (6) \quad \#$$

It is clear from Theorem (2.1) that, in general, there does not exist a unique design measure corresponding to each $\bar{M}$. Further, even if $\bar{M}(\xi) \in M$, it does not follow necessarily that $\xi \in \Xi_1$.

* Theorem 3

The moment space $M^{(p)}$ is a hyperplane in R^p iff the property $\bar{M}(\xi) \in M$ implies that $\xi \in \Xi_1$.

Proof

Necessity. Let $\underline{x}^0 \in M^{(p)}$, a hyperplane in R^p. Then $\underline{x}^0 = L \cap M^{(p)}$ where L is the ray of $M_c^{(p)}$ passing through $\underline{x}^0$.

Assume that $\underline{x}^0$ represents $\bar{M}(\xi^0) \in M$ where $\xi^0 \notin \Xi_1$, i.e. $\int_0^\pi d\xi^0(\omega) = \alpha \neq 1$. Then $\alpha^{-1}\xi^0 \in \Xi_1$ and $\bar{M}(\alpha^{-1}\xi^0) \subset M$. The point $\underline{x}^1$ representing $\bar{M}(\alpha^{-1}\xi^0)$ is distinct from $\underline{x}^0$ and lies in both L and $M^{(p)}$ i.e.

$$\underline{x}^0 \neq \underline{x}^1 \in L \cap M^{(p)} = \underline{x}^0$$

This is a contradiction. Therefore $\xi^0 \in \Xi_1$.

Sufficiency. Assume that $\bar{M}(\xi) \in M$ implies that $\xi \in \Xi_1$ and further assume that $M^{(p)}$ is not a hyperplane in R^p. Then $\xi^0 \in \Xi_1$ can be chosen so that $\bar{M}(\xi^0) \in$ Int M. Then $\bar{M}(\xi^\alpha) \in$ Int M where $\xi^\alpha \triangleq (1+\alpha)\xi^0$ for some $\alpha > 0$. However $\xi^\alpha \notin \Xi_1$. Contradiction.   q.e.d. #

Theorem 2.1 indicates that only $\xi \in \mathcal{D}_1$ need be considered in the design problem. Of particular interest are the designs with lowest index. The following definitions are of use and then it is shown in the next section that the index is closely related to the geometry of $M_c^{(p)}$.

<u>Definition 1</u>

If $\underline{x} \in M^{(p)}$, let $\mathcal{D}(\underline{x})$ denote the set of *discrete representations* of $\underline{x}$, i.e. the set of ξ such that $\underline{x} = \int_0^\pi \underline{v}(\omega) \, d\xi(\omega)$, where $\xi \in \mathcal{D}_1$.

<u>Definition 2</u>

Define the index $I(\underline{x})$ of a point $\underline{x}$ as

$$I(\underline{x}) = \min_{\xi \in \mathcal{D}(\underline{x})} I(\xi) \tag{7}$$

Clearly, if $\underline{x} = \underline{x}_1 + \underline{x}_2$ where $\underline{x}, \underline{x}_1, \underline{x}_2 \in M_c^{(p)}$, then $I(\underline{x}) \leq I(\underline{x}_1) + I(\underline{x}_2)$.

<u>Remark 1</u>

If $I(\underline{x}) = \ell$, then $\underline{x}$ has a representation of the form

$$\underline{x} = \sum_{i=1}^{k} \lambda_i \underline{v}(\omega_i) \tag{8}$$

where $k = \ell$ or $\ell+1$ and $\lambda_i > 0$, $i = 1, 2, \ldots, k$. The representation is said to have the spectrum $\omega_1, \ldots, \omega_k$ with corresponding weights $\lambda_1, \ldots, \lambda_k$.

* <u>Result 2</u>

If $\underline{x}$ represents a singular $\bar{M}$, then

$$I(\underline{x}) = I(\xi) = \tfrac{1}{2} \, \text{rank} \, (\bar{M}) \quad \forall \, \xi \in \mathcal{D}(\underline{x})$$

3.5/6

55

Proof

This follows directly from the corollary to Theorem (2.8.2).   #

3.6   CANONICAL REPRESENTATIONS

The following two theorems relate the boundary and interior of $M_c^{(p)}$ to the indices of points lying in them.

Theorem 1 [K8, p. 42]

A non-zero vector $\underline{x}^0$ lies on the boundary of $M_c^{(p)}$ iff $I(\underline{x}^0) < p/2$.

Proof

See Main Appendix.   #

Theorem 2 [K8, p. 44]

Let $\underline{x} \in \text{Int } M_c^{(p)}$. For each ω^* in $[0,\pi]$, there exists a representation

$$\underline{x} = \sum_{i=1}^{\ell} \lambda_i \underline{v}(\omega_i) \tag{1}$$

where $\lambda_i > 0$, $i = 1, \ldots, \ell$, with design index $p/2$ or $(p+1)/2$ and such that $\omega_i = \omega^*$ for some i.

Proof

See Main Appendix.   Result (4.1) is needed.   #

From Theorem 1 and Theorem (2.8.1) it follows that:

* Result 1

A point $\underline{x} \in M_c^{(p)}$ corresponds to a nonsingular information matrix iff $\underline{x} \in \text{Int } M_c^{(p)}$.

* Result 2

If the design measure ξ^* is Φ-optimal, then

$$\bar{M}(\xi^*) \in (\text{Int } M_c) \cap (\text{Bd } M)$$ #

Definition 1

Let $\underline{x} \in \text{Int } M_c^{(p)}$. A representation of $\underline{x}$ of index $\leq (p+1)/2$ is called *canonical* and any representation of index $p/2$ is called *principal*. A canonical or principal representation is further designated by the term *upper* if it involves the end point π and the term *lower* if it does not.

This leads to the following scheme for canonical representations.

Case I: p even

    (a) Upper canonical : $\frac{1}{2}p + 1$ frequencies including π

    (b) Lower canonical : $\frac{1}{2}p + 1$ frequencies including 0

    (c) Upper principal : $\frac{1}{2}p + 1$ frequencies including 0 and π (2)

    (d) Lower principal : $\frac{1}{2}p$ frequencies in $(0,\pi)$

Case II: p odd

    (a) Upper canonical : $\frac{1}{2}(p+1)+1$ frequencies including 0 and π

    (b) Lower canonical : $\frac{1}{2}(p+1)$ frequencies in $(0,\pi)$

    (c) Upper principal : $\frac{1}{2}(p+1)$ frequencies including π (3)

    (d) Lower principal : $\frac{1}{2}(p+1)$ frequencies including 0

Some important properties of canonical representations are brought together in the following theorem.

Theorem 3 [K8, pp. 47-49]

Let $\underline{x}^0 \in \text{Int } M_c^{(p)}$. Then

(i) $\underline{x}^0$ has precisely two principal representations. The spectra of these representations strictly interlace.

(ii) For any ω^* in $(0,\pi)$, there exists a unique canonical representation of $\underline{x}^0$ involving ω^*. For special choices of ω^*, the canonical representation happens to be a principal representation.   #

The above considerations show that any information matrix can be generated by using input designs composed of $[\frac{p+1}{2}]$ frequencies. This includes the elements of M. However this still does not necessarily provide a useful solution to the input design problem. It may be that no canonical representation of the Φ-optimal $\bar{M}*$ satisfies the input power constraint and this must be investigated further.

* Result 3

If $\bar{M}*$ is Φ-optimal in M and $\bar{M}* = \bar{M}(\xi)$ then

$$\int_0^\pi d\xi(\omega) \geq 1$$

Proof

If $\int_0^\pi d\xi(\omega) = \alpha < 1$, then $\alpha^{-1}\xi \in \Xi_1$ and $\bar{M}(\alpha^{-1}\xi) = \alpha^{-1}\bar{M}* \in M$. But $\alpha^{-1}\bar{M}* > \bar{M}*$. Therefore $\Phi[\alpha^{-1}\bar{M}*] < \Phi[\bar{M}*]$, contradicting the Φ-optimality of $\bar{M}*$. q.e.d. #

The main result is embodied in the following theorem.

* Theorem 4

If $q = 0$, $m \geq n+r$ and $\underline{x}^0 \in M^{(p)} \cap \text{Int } M_c^{(p)}$ then, for any $\omega*$ in $(0,\pi)$, there exists a unique canonical representation ξ^0 of $\underline{x}^0$ involving $\omega*$ and such that $\xi^0 \in \Xi_1$.

Proof

If $q = 0$, $m \geq n+r$ then $M^{(p)}$ is a hyperplane (theorem 5.2) and any ray of $M_c^{(p)}$ cuts $M^{(p)}$ in a unique point. Let $\underline{x}^0 \in M^{(p)} \cap \text{Int } M_c^{(p)}$ and let $\underline{x}* \in \text{Int } M_c^{(p)}$ and lie on the ray through $\underline{x}^0$. Then, by Theorem 3(ii) there exists a unique canonical representation of $\underline{x}*$ involving an arbitrary $\omega*$ in $(0,\pi)$. For some $\omega*$ let the total weight of the representation be μ. Then the point $\mu^{-1}\underline{x}^*$ has the same spectrum as $\underline{x}^*$ but with total weight unity. Therefore $\mu^{-1}\underline{x}^* \in M^{(p)}$. It follows that $\underline{x}^0 = \mu^{-1}\underline{x}^*$.   q.e.d. #

Clearly $\underline{x}^0$ may be chosen as the point corresponding to Φ-optimality where Φ can be any mapping from R^p into R (provided Φ-optimality excludes singular information matrices).

For the model (5.6), i.e. $y_k = z^{-s}B(z^{-1})u_k + e_k$, Φ-optimality can be achieved with a design comprising either $(m+1)/2$ frequencies (m odd) or $(m+2)/2$ frequencies (m even) without including 0 or π. For m even one of the frequencies can be chosen arbitrarily.

Theorem 4 gives a set of sufficient conditions (q = 0, m $\geq$ n+r) for a Φ-optimal canonical design to exist in Ξ_1. Other conditions are derived in the next section.

3.7 AN UPPER BOUND ON $I(\xi^*)$

In general, a point of $M^{(p)}$ will have an infinite number of representations. In some circumstances an upper bound can be placed on the index of those designs representing the point.

* ### Theorem 1

If $M^{(p)}$ is not a hyperplane in R^p and $\underline{x}^0 \in (Bd\,M^{(p)}) \cap (Int\,M_c^{(p)})$, then every design measure ξ in Ξ_1 representing $\underline{x}^0$ is discrete and satisfies the inequality

$$I(\xi) \leq \tfrac{1}{2}\max(p+q-1, 2n+r) \qquad (1)$$

Proof

If $M^{(p)}$ is not a hyperplane in R^p and $\underline{x}^0 \in Bd\,M^{(p)}$ then there exists a hyperplane supporting $M^{(p)}$ and passing through $\underline{x}^0$, i.e.

$\exists$ real numbers $\alpha_1, \ldots, \alpha_p$, d such that at least one α_i is non-zero,

d is non-zero if $\underline{x}^0 \in \text{Int } M_c^{(p)}$ and

$$\sum_{i=1}^{p} \alpha_i x_i \leq d \quad \forall \; \underline{x} \in M^{(p)} \tag{2}$$

$$\sum_{i=1}^{p} \alpha_i x_i^0 = d \tag{3}$$

where

$$x_i = \int_0^{\pi} f(\omega) \cos^{i-1} \omega \, d\xi(\omega), \quad \xi \in \Xi_1 \tag{4}$$

Substituting from (4) into (2)

$$\int_0^{\pi} \{ \sum_{i=1}^{p} \alpha_i f(\omega) \cos^{i-1} \omega - d \} d\xi(\omega) \leq 0 \quad \forall \; \xi \in \Xi_1 \tag{5}$$

and equality holds in (5) for $\xi = \xi_0$, a design measure in Ξ_1

representing $\underline{x}^0$.

From (5) it follows that

$$f(\omega) \sum_{i=1}^{p} \alpha_i \cos^{i-1} \omega \leq d \quad \forall \; \omega \in [0, \pi] \tag{6}$$

and that frequencies in the spectrum of ξ^0 must belong to the set

of real roots of

$$f(\omega) \sum_{i=1}^{p} \alpha_i \cos^{i-1} \omega = d \tag{7}$$

i.e.

$$\frac{pol_q(t)}{pol_{2n+r}(t)} pol_{p-1}(t) = d \text{ where } t = \cos \omega \in [-1,1]$$

i.e.

$$\text{pol}_\gamma(t) \equiv \text{pol}_q(t)\text{pol}_{p-1}(t) - d\ \text{pol}_{2n+r}(t) = 0 \tag{8}$$

where

$$\gamma = \max\ (p+q-1, 2n+r)$$

From (6) it follows that the real roots of (8) in $[-1,1]$ must be of even multiplicity, except possibly for $t = \pm 1$. Figure 1 shows the case where ξ^0 is a representation with spectrum $(\omega_1, \omega_2, \ldots, \omega_\ell)$ where $\omega_1 = 0$, $\omega_\ell < \pi$.

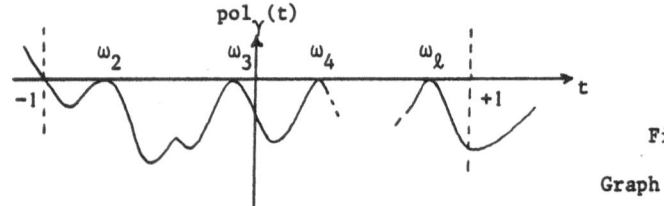

Figure 1

Graph of $\text{pol}_\gamma(t)$

Let N_0 be the number of real roots of $\text{pol}_\gamma(t)$ lying in $[-1,1]$ and counting multiplicities. Then

$$\gamma \geq N_0 \geq 2(\ell-1) + 1$$

and

$$I(\xi_0) = (\ell-1) + \tfrac{1}{2}$$

The result follows immediately and it is easily established that, in every case, $\gamma \geq 2I(\xi_0)$ leading to (1).   #

Remark 1

Some modification to Theorem 1 is necessary if the polynomials A, C have common roots. For example, if there are k roots in common then $f(\omega) = pol_{q-k}(t)/pol_{2n+r-k}(t)$ leading to $\gamma = \max (p+q-1,2n+r) - k$.

Condition (1) is then replaced by

$$I(\xi) \leq \tfrac{1}{2} \max (p+q-1,2n+r) - \tfrac{1}{2}k \qquad\qquad (1a)$$

Remark 2

If $M^{(p)}$ is not a hyperplane, then $\gamma \geq p$.

Clearly no absolutely continuous design measure can satisfy condition (1). This implies that an input sequence that is a realisation of a stochastic process (subject to remark 2.7.1) cannot correspond to the point $\underline{x}^0$ of Theorem 1. In particular, if $\underline{x}^0$ is the Φ-optimum, then:

* Corollary 1

If a stochastic Φ-optimal design exists then $M^{(p)}$ is a hyperplane.

In particular, Levin [L2] shows that the L-optimal input for weighting function models (5.6) has an impulsive autocorrelation, a property of discrete white noise. This corresponds to an optimal design measure $\xi^* \in \Xi_1$ given by

$$\xi^*(\omega) = \pi^{-1}\omega, \quad \omega \in [0,\pi] \qquad\qquad (9)$$

Of course, Theorem (3.2.1) ensures that there exists in each case an optimal design with a discrete spectrum yielding an input with an impulsive autocorrelation (up to the required number of shifts).

It is clear that the upper bound in (1) may be far larger than $(p+1)/2$ for high order noise models. Theorem 2 lays down sufficient conditions for ruling out non-canonical representations.

* Theorem 2

If $\underline{x}^0 \in (\text{Bd} M^{(p)}) \cap (\text{Int } M_c^{(p)})$ and $M^{(p)}$ is not a hyperplane in R^p, then a sufficient condition that every representation in Ξ_1. of $\underline{x}^0$ is canonical is that

either (i) q = 0; r = m - n + 1 or m - n + 2

or (ii) q = 1 or 2; r ≤ m - n + 2

$$(10)$$

Further, if

either (iii) q = 0; r = m - n + 1

or (iv) q = 1; r ≤ m - n + 1

$$(11)$$

then only principal representations of $\underline{x}^0$ are possible.

Proof

If the upper bound in (1) does not exceed (p+1)/2 then no non-canonical representation of $\underline{x}^0$ can exist. This requires that $\gamma \le p + 1$ and leads to (10). Similarly, the condition $\gamma = p$ leads to (11). #

If a discrete-time system is derived by sampling a continuous-time output (see figure 2) where H(s) is assumed to be a proper rational transfer function plus a time delay τ, then, in general, m=n-1 if τ is an integer multiple of the sampling interval and m=n otherwise. Assuming that the sampled output is corrupted by discrete white noise, i.e, q = 0 = r, then in the first case the polynomial orders satisfy condition (iii) of Theorem 2 and in the second case the hyperplane conditions (5.3) are satisfied. Therefore, in either case, only principle representations need be considered.

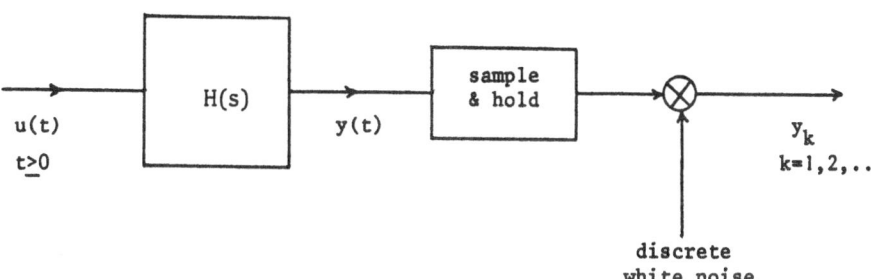

Figure 2

The hyperplane condition (5.3) and the sets of conditions (10) and (11) are each sufficient to ensure that there exist Φ-optimal canonical designs satisfying the input power constraint. These conditions are not necessary and, unfortunately, such conditions have not yet been established.

Javaherian [J1] has run a considerable number of computational examples, all of which achieve D-optimality with $\lceil (p+1)/2 \rceil$ frequencies. Using the above geometrical approach, however, counter-examples can be constructed to his tentative hypothesis that this result is generally true (see Appendix B).

3.8   OUTPUT POWER CONSTRAINT

In this section, the approach developed earlier in the chapter is extended to the case of output power constraint. Following the discussion of section (2.7), the information matrix can be written as

$$\bar{M} = \text{Re} \int_0^{\pi} \tilde{h}(e^{j\omega}) \tilde{h}^*(e^{j\omega}) d\eta(\omega) \tag{1}$$

where

$$\tilde{h}_i(z) = \frac{C}{DA}(z) z^{s+i} \qquad i = 1, \ldots, n$$

$$= -\frac{C}{DB}(z) z^{s+i-n-1} \qquad i = n+1, \ldots, p \tag{2}$$

and

$$\int_0^\pi d\eta(\omega) = 1 \tag{3}$$

The design measure $\eta(\omega)$ corresponds to the power distribution of the input-dependent component of the output.

Notation

As in (1), the same symbols are used as in the input power constraint analysis with $\sim$ over each one.

The form of (1), (2) and (3) is almost exactly the same as in the input power constraint case. The only change is that $\frac{C}{DA^2}(z)$ is replaced by $\frac{C}{DAB}(z)$. This leads to $\{\tilde{v}_i\}_1^p$ as a T-system on $[0,\pi]$ where

$$\tilde{v}_i(\omega) = \tilde{f}(\omega) \cos^{i-1}\omega \qquad i = 1, \ldots, p \tag{4}$$

$$\tilde{f}(\omega) = \frac{C}{DAB}(e^{j\omega}) \frac{C}{DAB}(e^{-j\omega}) \tag{5}$$

The following theorems are of particular interest.

* Theorem 1

The moment space $\tilde{M}^{(p)}$ is a hyperplane in R^p iff the system output is corrupted by discrete white noise i.e. $q = 0 = r$.

Proof

See Theorem (5.2).

Remark 1

Note that the condition placed on the orders of the process polynomials is far less restrictive than in the case of input power constraint. Here no restriction is placed on the system polynomials A, B at all.   #

3.8

65

* **Theorem 2**

If $q = 0 = r$ and $\underline{x}^0 \in \widetilde{M}^{(p)} \cap \text{Int } \widetilde{M}_c^{(p)}$ then, for any ω^* in $(0,\pi)$, there exists a unique canonical representation η^0 of $\underline{x}^0$ involving ω^* and such that $\eta^0 \in \Xi_1$.

Proof

See Theorem (6.4). #

Remark 2

Note that if $q = 0 = r$ and $m \geq n$, then both $M^{(p)}$ and $\widetilde{M}^{(p)}$ are hyperplanes in R^p.

* **Theorem 3**

If $\widetilde{M}^{(p)}$ is not a hyperplane in R^p and $\underline{x}^0 \in (\text{Bd}\vec{\widetilde{M}}^{(p)}) \cap (\text{Int } \widetilde{M}_c^{(p)})$, then every design measure η in Ξ_1 representing $\underline{x}^0$ is discrete and satisfies the inequality

$$I(\eta) \leq \tfrac{1}{2}(p-1) + \tfrac{1}{2} \max (r,q) \tag{6}$$

Proof

See Theorem (7.1) #

* **Theorem 4**

If $\underline{x}^0 \in (\text{Bd }\widetilde{M}^{(p)}) \cap (\text{Int } \widetilde{M}_c^{(p)})$ and $\widetilde{M}^{(p)}$ is not a hyperplane in R^p, a sufficient condition that every representation in Ξ_1 of $\underline{x}^0$ is canonical is that

$$\max (r,q) \leq 2 \tag{7}$$

If $\max (r,q) = 1$, then only principal representations of $\underline{x}^0$ are possible.

Proof

See Theorem (7.2).   #

It is clear that the upper bounds in (7.1) and (6) will (in many cases) allow a reduction of the number of search variables from 2p to max (p+q-1,2n+r) or p - 1 + max (r,q) respectively when seeking a Φ-optimal design.

3.9   CONCLUDING REMARKS

This chapter has developed a geometrical approach to the problem of optimal input design based on Tchebycheff system theory. Sufficient conditions have been derived for the existence of certain minimal representations of Φ-optimal designs, leading to a reduction of the order of the optimization problem. In the next chapter it is shown that the assumption of D-optimality leads to further knowledge of optimal designs.

Appendix A

EXAMPLES

EXAMPLE 1

Consider the case $m = 1$, $n = 0$, $r = 1$, $q = 1$, i.e. a model with two system parameters and of the form

$$y_k = (a+bz^{-1})u_k + \frac{c+dz^{-1}}{1+fz^{-1}} e_k, \quad k = 1, \ldots, N \qquad (1)$$

where

$$e_k \sim N(0,1)$$

Let ξ_ω denote the single frequency design in Ξ_1 corresponding to the frequency ω. Then, from (2.6.14) and (2.6.15),

$$\bar{M}(\xi_\omega) = \frac{1+f^2+2f \cos \omega}{c^2+d^2+2cd \cos \omega} \begin{bmatrix} 1 & \cos \omega \\ \cos \omega & 1 \end{bmatrix} \qquad (2)$$

$$= \sum_{i=1}^{2} x_i L_i = \begin{bmatrix} x_1 & x_2 \\ x_2 & x_1 \end{bmatrix}$$

where

$$L_1 = \begin{bmatrix} 1 & 0 \\ 0 & 1 \end{bmatrix}, \quad L_2 = \begin{bmatrix} 0 & 1 \\ 1 & 0 \end{bmatrix}$$

and

$$x_1 = \frac{1+f^2+2f \cos \omega}{c^2+d^2+2cd \cos \omega} \quad ; \quad x_2 = x_1 \cos \omega \qquad (3)$$

The convex set $M^{(2)}$ is the shaded portion in figure 1, where it is assumed $c = 1$, $d = 0.3$. $f = 0.5$.

The cone $M_c^{(2)}$ is generated by rays from the origin through every point of $M^{(2)}$.

The curved boundary of $M^{(2)}$ is the trajectory corresponding to single frequency designs on $[0,\pi]$. In particular, the design which maximises det $\bar{M}(\xi)$ lies on this boundary at $\omega* = 1.38$ where $x_1^* = 1.22$, $x_2^* = 0.23$. From (2) and (3)

$$\bar{M}(\xi_\omega)^{-1} = \begin{bmatrix} x_1 & x_2 \\ x_2 & x_1 \end{bmatrix}^{-1} = (x_1^2 - x_2^2)^{-1} \begin{bmatrix} x_1 & -x_2 \\ -x_2 & x_1 \end{bmatrix}$$

Therefore, for the optimal design

$$\text{var } \hat{a} = \text{var } \hat{b} = 0.85/N; \quad \text{det } \bar{M} = 1.38 \tag{4}$$

For white noise input:

$$\bar{M} = \frac{1}{c^2(c^2-d^2)} \begin{bmatrix} c^2(1+f^2)-2cfd & (cf-d)(c-fd) \\ (cf-d)(c-fd) & c(1+f^2)-2cfd \end{bmatrix}$$

$$= \begin{bmatrix} 1.04 & 0.19 \\ 0.19 & 1.04 \end{bmatrix}$$

Then

$$\text{var } \hat{a} = \text{var } \hat{b} = 1.00/N; \quad \text{det } \bar{M} = 1.05 \tag{5}$$

From the figure it is clear that the point $(1.04, 0.19)$, corresponding to white noise input, can be expressed as a convex combination of two single frequency designs in an infinite number of ways.

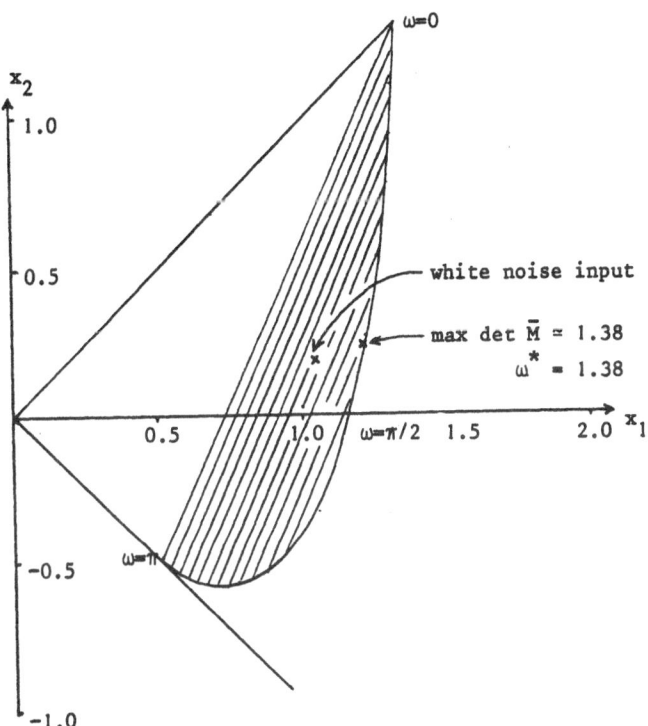

Figure 1

For example,

$$(1.04, 0.19) = 0.37 \ (1.33, 1.33) + 0.63 \ (0.87, -0.48)$$

$$\downarrow \qquad \downarrow \qquad \downarrow \qquad \downarrow$$

$$\lambda_1 \qquad \omega_1 = 0 \qquad \lambda_2 \qquad \omega_2 = 2.16$$

yields a discrete design with index 3/2 equivalent to white noise.

EXAMPLE 2

Consider the first example with $f \equiv 0$, i.e.

$$y_k = (a + bz^{-1})u_k + (c + dz^{-1})e_k, \quad k = 1, \dots, N \qquad (6)$$

The model orders now satisfy the hyperplane condition. From (3)

$$x_1 = (c^2+d^2+2cd \cos \omega)^{-1}; \quad x_2 = (c^2+d^2+2cd \cos \omega)^{-1} \cos \omega$$

$$(c^2+d^2)x_1 + 2cdx_2 = 1 \tag{7}$$

This is a straight line connecting the points $(1,1)/(c+d)^2$ and $(1,-1)/(c-d)^2$ corresponding to $\omega = 0$ and $\omega = \pi$ respectively, i.e. all points of $M^{(2)}$ can be realized by single frequency designs. In particular $\cos \omega = -2cd/(c^2+d^2)$ gives the maximum value for $\det \bar{M}$ of $(c^2-d^2)^{-2}$. This yields

$$\bar{M}^{-1} = \begin{bmatrix} c^2+d^2 & 2cd \\ 2cd & c^2+d^2 \end{bmatrix}$$

so that

$$\text{var } \hat{a} = \text{var } \hat{b} = (c^2+d^2)/N \tag{8}$$

For white noise input, $\det \bar{M} = [c^2(c^2-d^2)]^{-1}$ and

$$\bar{M}^{-1} = c^2 \begin{bmatrix} 1 & d/c \\ d/c & 1 \end{bmatrix}; \quad \text{var } \hat{a} = \text{var } \hat{b} = c^2/N \tag{9}$$

However, the single frequency design $\omega = \cos^{-1}(-d/c)$ yields the same result.

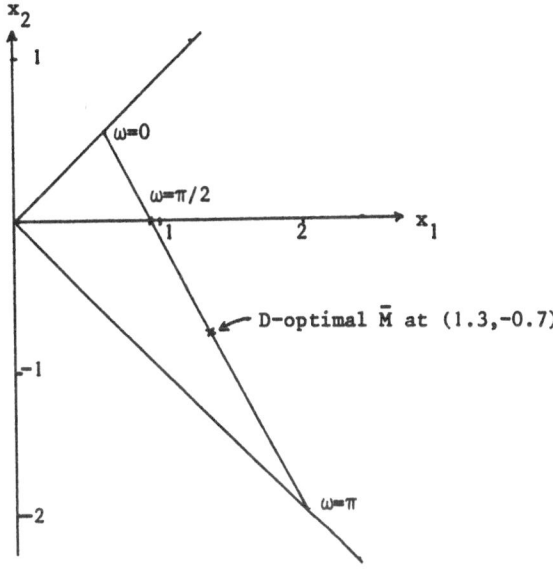

Figure 2: The case f = 0

EXAMPLE 3

Consider the model

$$Y_k = \frac{b_0 + b_1 z^{-1}}{1 + a_1 z^{-1}} u_k + e_k, \quad k = 1, \ldots, N \tag{10}$$

In this case p = 3 and the hyperplane condition is satisfied, i.e. $M^{(3)}$ lies in a plane in Euclidean 3-space. The set $M^{(3)}$ is represented in figure 3. The curved boundary is the trajectory of single frequency design points.

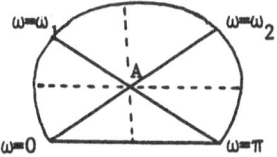

Figure 3

Let A be an arbitrary point in the relative interior of $M^{(3)}$.
Then A has two interlacing principal representations with spectra
$(0,\omega_2)$ and (ω_1,π) respectively, as shown in the figure. Also let
ω^* be an arbitrary frequency in $[0,\pi]$.

If $\omega^* \in [0,\omega_1] \cup [\omega_2,\pi]$, then A has a unique two frequency
representation involving ω^*. However, if $\omega^* \in (\omega_1,\omega_2)$ then the
three frequencies 0, ω^*, π must be used to provide a unique canonical
representation of A. The principal representations correspond to
$\omega^* = \omega_1$ or ω_2. These properties are clear from the geometry of
figure 3.

If $b_0 = 1$, $b_1 = 0.3$, $a_1 = 0.5$, $e_k \sim N(0,0.01)$ and A is the
D-optimal design point, then A has the following representations:

	Frequency	Power Proportion	
Lower Principal :	0	1/3	
	2.76 $(=\omega_2)$	2/3	
Upper Principal :	2.09 $(=\omega_1)$	2/3	
	π	1/3	
Canonical :	1.00 $(=\omega^*)$	0.36	$\omega^* \in [0,\omega_1]$
	2.78	0.64	
	1.88	0.54	
	2.90 $(=\omega^*)$	0.46	$\omega^* \in [\omega_2,\pi]$
	0	0.15	
	2.30 $(=\omega^*)$	0.55	$\omega^* \in (\omega_1,\omega_2)$
	π	0.30	

These designs correspond to the cost value

$$\det \bar{\bar{M}}^{-1} = 4.45 \times 10^{-6} \tag{11}$$

where $\bar{\bar{M}}^{-1}$ is given by

$$\bar{\bar{M}}^{-1} = \begin{bmatrix} 0.079 & 0.007 & 0.104 \\ 0.007 & 0.013 & 0.019 \\ 0.104 & 0.019 & 0.149 \end{bmatrix} \tag{12}$$

If a search is carried out over the set of all two-frequency designs for which one of the frequencies is fixed at $\omega^* = 2.30$, then the best design is

Frequency	Power Proportion
0	1/3
2.30	2/3

giving a cost value of $\det \bar{\bar{M}}^{-1} = 1.6 \times 10^{-5}$, where

$$\bar{\bar{M}}^{-1} = \begin{bmatrix} 0.207 & 0.037 & 0.232 \\ 0.037 & 0.016 & 0.046 \\ 0.232 & 0.046 & 0.269 \end{bmatrix}$$

Comparison with (11) and (12) shows that this is far from D-optimal and bears out the preliminary remarks on the geometry of $M^{(3)}$.

For white noise input, $\det \bar{\bar{M}}^{-1} = 7.90 \times 10^{-6}$ and var $\hat{a}_1 = 0.106/N$, var $\hat{b}_0 = 0.010/N$, var $\hat{b}_1 = 0.146/N$. Comparison with the diagonal elements in (12) shows that there is a slight improvement in the estimation accuracy of parameters b_0, b_1 at the expense of parameter a_1.

Appendix B

COUNTEREXAMPLES TO JAVAHERIAN'S CONJECTURE

Javaherian [J1] has computed ℓ-frequency designs that maximise det $\bar{M}$ for process models of the form (2.2.6). For all computed cases, employing either input or output power constraints, it is found that D-optimal designs are achieved provided $\ell \geq [(p+1)/2]$. This indicates that optimality may be achieved in general by searching over those discrete designs whose spectra just satisfy the minimum conditions for persistent excitation.

However, it can be shown that this conjecture does not hold and a simple counterexample in the case of two system parameters indicates a general method for constructing such counterexamples for arbitrary p.

EXAMPLE 1

Consider the model

$$y_k = (1+0.3z^{-1})u_k + \frac{1}{1-0.06z^{-4}} e_k \tag{1}$$

The set $M^{(2)}$ is the convex hull of the single frequency curve, a point (x_1, x_2) of which is given by

$$\begin{aligned} x_1 &= f(\omega) \\ x_2 &= f(\omega) \cos \omega \end{aligned} \qquad \omega \in [0,\pi] \tag{2}$$

where $f(\omega) = 1.0036 - 0.12 \cos 4\omega$.

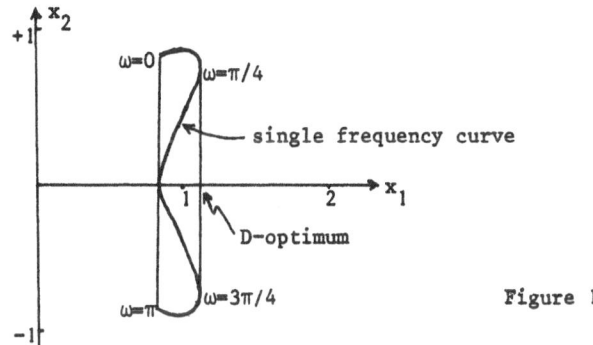

single frequency curve

D-optimum

Figure 1

From figure 1 it can be seen that, unlike example A1, the single frequency curve does not lie entirely in the boundary of $M^{(2)}$. Points on the straight line portions of $\mathrm{Bd}\, M^{(2)}$ cannot be represented by single frequency designs in Ξ_1, but require at least two frequencies: $(0,\pi)$ or $(\pi/4, 3\pi/4)$.

Computation reveals that the D-optimal design has a representation with frequencies $\pi/4$, $3\pi/4$ and equal power proportions, yielding the cost function value

$$\det \bar{\bar{M}}^{-1} = 0.80$$

where

$$\bar{\bar{M}}^{-1} = \begin{bmatrix} 0.90 & 0 \\ 0 & 0.90 \end{bmatrix}$$

The smallest cost value achievable with a single frequency design is

$$\det \bar{\bar{M}}^{-1} = 1.16$$

where

$$\overline{M}^{-1} = \begin{bmatrix} 1.18 & \underline{+}0.47 \\ \underline{+}0.47 & 1.18 \end{bmatrix} \qquad (3)$$

The positive sign in (3) corresponds to the single frequency $\omega_0 = 1.16$ and the negative sign to the frequency $\pi - \omega_0$.

In figure 1, D-optimality is achieved at (1.12,0), however, for other choices of Φ, Φ-optimality may occur away from the x_1-axis and have a single frequency representation ω_0 where $\omega_0 \in [0,\pi/4]$, say.

The structure of $M^{(2)}$ in example 1 has some features that suggest a general approach to the construction of counterexamples. Roughly, the impossibility of representing the D-optimum in example 1 by a single frequency design in Ξ_1 is due to the fact that a part of Bd $M^{(2)}$ away from the origin is a 2-dimensional simplex symmetrical about the x_1-axis, the axis of the cone $M_c^{(2)}$. This suggests that counter-examples can be constructed by choosing the process model so that, for given p, part of Bd $M^{(p)}$ away from the origin is a k-dimensional simplex where $k > [(p+1)/2]$. In particular, this occurs if the distance from the origin of points on the single frequency curve exhibits k equal maxima.

Let $t = \cos \omega$, $\psi(t) = f(\omega)$ and $d(t)$ be the distance from the origin of the point on the single frequency curve parametrised by t. Then

$$d(t) = \psi(t) \cdot (\sum_{i=0}^{p-1} t^{2i})^{\frac{1}{2}}, \qquad t \in [-1,1] \qquad (4)$$

where

$$\psi(\cos \omega) = \frac{C}{DA^2}(e^{j\omega}) \frac{C}{DA^2}(e^{-j\omega}) \underline{\Delta} \frac{\psi_1(t)}{\psi_2(t)} \qquad (5)$$

and ψ_1, ψ_2 are the numerator and denominator polynomials respectively of $\psi(t)$.

Proceed as follows:

    (i)   Choose $A(z)$, $D(z)$ so that ψ_2 is known.

    (ii)   Select any function for $d(t)$ with the required behaviour, e.g. $\tilde{d}(t) \underline{\Delta} 1 + 2 \sin^2 \pi \ell t$ is positive and has 2ℓ equal maxima in $(-1,1)$.

    (iii)   Derive the values of the polynomial coefficients in $\psi_1(t)$ by fitting it by least squares to

$$\psi_2(t)\tilde{d}(t) \; (\sum_{i=0}^{p-1} t^{2i})^{-\frac{1}{2}} \tag{6}$$

    (iv)   Factorize $\psi_1(\cos \omega)$ by solving the equation

$$c(e^{j\omega})c(e^{-j\omega}) = \psi_1(\cos \omega) \tag{7}$$

so that $c(z)$ does not vanish on the closed unit disc.

    This procedure always yields the required structure for $M^{(p)}$ provided r, the order of $C(z)$, is chosen large enough.

EXAMPLE 2

    Consider the model

$$y_k = B(z^{-1})u_k + \frac{1}{C(z^{-1})} e_k \tag{8}$$

Using the above procedure in the case $p = m+1 = 3$, the set $M^{(3)}$ can be constructed with a triangular face away from the origin. This is achieved by the choice:

$$B(z) = 1 + 0.5z + 0.06z^2$$

$$C(z) = 1 - 0.23z^2 - 0.08z^4 - 0.07z^6 \tag{9}$$

This model leads to a D-optimal three-frequency design:

Frequency	Power Proportion
0.67	0.327
$\pi/2$	0.346
$\pi-0.67$	0.327

yielding a cost value

$$\det \bar{M}^{-1} = 0.510$$

The best two frequency design is $(\pi/4, 3\pi/4)$ with equal power proportions and yields a cost value of 0.590.

The same behaviour occurs for A-optimality and also for L-optimality with $W = \text{diag}(10,1,1)$.

These results show that, in general, in order to achieve Φ-optimality it may be necessary to search over the set of p-frequency representations and that only under certain conditions can Payne's result (Theorem 3.1) be strengthened.

Chapter 4

D-OPTIMAL DESIGNS

4.1   INTRODUCTION

In this chapter attention is confined to designs that are optimal under the determinant criterion. It is shown that, if principal D-optimal designs exist, their power proportions are known (Section 2) and in some cases the optimal design can be completely determined analytically (Section 3).

The main sections of the chapter are concerned with possible sequential design procedures converging to a D-optimal design (Sections 5, 7, 9). The basic framework rests on the Kiefer-Wolfowitz equivalence theorem (Section 4) and a useful extension (Section 8). Global convergence is proved for a particular class of algorithms (Section 9) and the rates of convergence of some of these algorithms are compared in an appendix. Some further properties of D-optimal designs arise from the sequential approach (Section 6). A sequential 'rounding-off' procedure is proposed in Section 10.

4.2   PRINCIPAL D-OPTIMAL DESIGNS

Consider the ℓ-frequency design $\{\lambda_i, \omega_i,\ i=1,\dots,\ell\}$ yielding the information matrix

$$\bar{M} = \mathrm{Re}\ \sum_{k=1}^{\ell} \lambda_k h(e^{j\omega_k})h^*(e^{j\omega_k}) \tag{1}$$

in the usual notation.

Introducing the sequence of vectors

$$g(\omega_k) = \sqrt{(-\frac{\lambda_k}{2})} h(e^{j\omega_k}) \qquad k = 1, 2, \ldots, \ell \qquad (2)$$

and the px2ℓ matrix F given by

$$F = [g(\omega_1)\bar{g}(\omega_1) g(\omega_2)\bar{g}(\omega_2) \ldots g(\omega_\ell)\bar{g}(\omega_\ell)] \qquad (3)$$

equation (1) can be written in the form

$$\bar{M} = FF* \qquad (4)$$

Forming the determinant of $\bar{M}$ yields

$$|\bar{M}| = |FF*| = \sum_i |F_i| \, |F_i^*| \qquad (5)$$

where F_i is a pxp matrix formed from F by deleting 2ℓ-p columns and
the summation is taken over the $_{2\ell}C_p$ terms generated in this way
(see Hadley [H2, p. 102]).

In general maximising $|\bar{M}|$ for arbitrary ℓ cannot be carried out
analytically. However, if it is known that a principal D-optimal
design exists (see Chapter 3) the following result allows a significant
reduction in the number of search variables.

* <u>Theorem 1</u>

Any principal D-optimal design in $\mathcal{D}_1$ with representation

$\{\lambda_i, \omega_i, \; i=1,2,\ldots,\ell\}$ has the power distribution

$$\lambda_i = 2/p \text{ if } \omega_i \in (0,\pi)$$
$$ = 1/p \text{ if } \omega_i = 0 \text{ or } \pi \qquad\qquad i = 1, 2, \ldots, \ell \qquad\qquad (6)$$

Proof

From (3.6.2) and (3.6.3) the following cases occur:

Case (i): p even, $2\ell = p$, $0 < \omega_1 < \ldots < \omega_\ell < \pi$. In this case F is a square matrix and, using (2) and (3)

$$|\bar{M}| = \prod_{k=1}^{\ell} \lambda_k^2 \; x \; (\text{term independent of } \lambda_1,\ldots,\lambda_\ell)$$

The maximum of $|\bar{M}|$ subject to the constraint

$$\sum_{i=1}^{\ell} \lambda_i = 1 \qquad\qquad (7)$$

occurs when $\lambda_1 = \lambda_2 = \ldots = \lambda_\ell$ and therefore

$$\lambda_i = 2/p, \quad i = 1, \ldots, \ell$$

Case (ii): p odd, $2\ell = p+1$, $\omega_1 = 0$ or $\omega_\ell = \pi$. From (5)

$$|\bar{M}| = \sum_{i=1}^{2\ell} |F_i||F_i^*| \qquad\qquad (8)$$

where F_i denotes F with the ith column deleted. Let F denote the $p \times 2\ell$ matrix given by

$$F = [h(e^{j\omega_1})h(e^{-j\omega_1}) \ \ldots \ h(e^{j\omega_\ell})h(e^{-j\omega_\ell})] \tag{9}$$

and let F_i denote F with the ith column deleted. Then

$$|F_{2k-1}| = 2^{-p/2}\{\prod_{i=1}^{\ell} \lambda_i\}\lambda_k^{-\frac{1}{2}}|F_{2k-1}|$$

$$k = 1, \ 2, \ \ldots, \ \ell \tag{10}$$

$$|F_{2k}| = 2^{-p/2}\{\prod_{i=1}^{\ell} \lambda_i\}\lambda_k^{-\frac{1}{2}}|F_{2k}|$$

Therefore, substituting in (8)

$$|\bar{M}| = \{\prod_{i=1}^{\ell} \lambda_i^2\} \sum_{k=1}^{\ell} \lambda_k^{-1}\tilde{F}_k \tag{11}$$

where

$$\tilde{F}_k = 2^{-p}\{|F_{2k-1}| \ |F_{2k-1}^*| + |F_{2k}| \ |F_{2k}^*|\}, \quad k = 1, \ 2, \ \ldots, \ \ell$$

To maximise $|\bar{M}|$ subject to (7), consider the augmented cost function

$$J = |\bar{M}| - \mu \sum_{i=1}^{\ell} \lambda_i \tag{12}$$

where μ is a Lagrange multiplier. Then a necessary condition for D-optimality is $\partial J/\partial\lambda_k = 0$, $k = 1, \ \ldots, \ \ell$ and this yields

$$\{\prod_{i=1}^{\ell} \lambda_i^2\}[2\sum_{j=1}^{\ell} \lambda_j^{-1}\tilde{F}_j - \lambda_k^{-1}\tilde{F}_k] - \mu\lambda_k = 0, \ k = 1, \ \ldots, \ \ell \tag{13}$$

Summing over k

$$\mu = p\{\prod_{i=1}^{\ell} \lambda_i^2\} \sum_{k=1}^{\ell} \lambda_k^{-1}\tilde{F}_k \tag{14}$$

and substituting back in (13)

$$\lambda_k (2-p\lambda_k) \sum_{j=1}^{\ell} \lambda_j^{-1} \tilde{F}_j = \tilde{F}_k, \quad k = 1, 2, \ldots, \ell \qquad (15)$$

For the lower principal representation, in this case, $\omega_1 = 0$ and this yields

$$\left| F_i \right| = 0, \quad i = 3, 4, \ldots, 2\ell$$

and

$$\tilde{F}_i = 0, \quad i = 2, 3, \ldots, \ell$$

leading to

$$\lambda_k = 2/p, \quad k = 2, 3, \ldots, \ell$$

and therefore

$$\lambda_1 = 1 - \sum_{i=2}^{\ell} \lambda_i = 1/p$$

For the upper principal representation, $\omega_\ell = \pi$ and

$$\lambda_k = 2/p, \quad k = 1, 2, \ldots, \ell-1; \quad \lambda_\ell = 1/p$$

__Case (iii)__: p even, $2\ell = p+2$, $0 = \omega_1 < \omega_2 < \ldots < \omega_\ell = \pi$. Let F_{ij} denote F with i^{th} and j^{th} columns deleted $(i < j)$; similarly for $\tilde{F}_{ij}$. Then

$$\left| \bar{M} \right| = \sum_{1 \leq i < j \leq 2\ell} \left| F_{ij} \right| \left| F^*_{ij} \right| \qquad (16)$$

where

$$\left| F_{ij} \right| = 2^{-p/2} \{ \prod_{s=1}^{\ell} \lambda_s \} \lambda_k^{-\frac{1}{2}} \lambda_t^{-\frac{1}{2}} \left| \tilde{F}_{ij} \right| \qquad (17)$$

and

$$i = 2k \text{ or } 2k-1, \quad j = 2t \text{ or } 2t-1, \quad i < j$$

$$k, t = 1, 2, \ldots, \ell$$

Define F_{ij}, F_{ij} as zero for $i \geq j$. Then

$$\sum_{1 \leq i < j \leq p+2} |F_{ij}||F_{ij}^*| = \sum_{k=1}^{\ell} \sum_{t=1}^{\ell} \tilde{F}_{kt} \tag{18}$$

where

$$\tilde{F}_{kt} = |F_{2k,2t}||F_{2k,2t}^*| + |F_{2k,2t-1}||F_{2k,2t-1}^*|$$

$$+ |F_{2k-1,2t}||F_{2k-1,2t}^*| + |F_{2k-1,2t-1}||F_{2k-1,2t-1}^*| \tag{19}$$

and

$$\tilde{F}_{kt} = 0 \text{ if } k > t$$

Substituting into (16)

$$|\bar{M}| = 2^{-p} \{ \prod_{s=1}^{\ell} \lambda_s^2 \} \sum_{k=1}^{\ell} \sum_{t=1}^{\ell} (\lambda_k \lambda_t)^{-1} \tilde{F}_{kt} \tag{20}$$

Maximising $|\bar{M}|$ subject to the power constraint (7) can be carried through as in case (ii). This leads to the following equations in place of (15):

$$\lambda_i (2-p\lambda_i) \sum_{k=1}^{\ell} \sum_{t=1}^{\ell} (\lambda_k \lambda_t)^{-1} \tilde{F}_{kt} = \sum_{t=i}^{\ell} \lambda_t^{-1} \tilde{F}_{it} + \sum_{k=1}^{i} \lambda_k^{-1} \tilde{F}_{ki}$$

$$i = 1, 2, \ldots, \ell \tag{21}$$

If $\omega_1 = 0$ and $\omega_\ell = \pi$, the only non-zero $|F_{ij}|$ occur for $i = 1, 2$
and $j = 2\ell-1, 2\ell$ and therefore $\tilde{F}_{ij}$ is zero unless $i = 1, j = \ell$. It
follows from (21) that

$$\lambda_1 = 1/p$$

$$\lambda_k = 2/p \qquad\qquad k = 2, 3, \ldots, \ell-1 \qquad\qquad (22)$$

$$\lambda_\ell = 1/p$$

Cases (i) - (iii) cover all possible principal representations and
the proof is complete.   #

Remark 1

Case (i) is well known, e.g. [K8, p. 332].

Remark 2

If $M^{(p)}$ is a hyperplane then Theorem 1 and Theorem (3.6.4) together
allow the search for a D-optimal design to be carried out in a
$[(p-1)/2]$-dimensional subspace of R^p. Only the $[(p-1)/2]$ frequencies
$\omega_2, \omega_3, \ldots, \omega_{[(p+1)/2]}$ remain undetermined in cases (ii) and (iii)
of Theorem 1.

In case (i), with some extra assumptions, it is possible to derive
an explicit expression for $|\bar{M}|$ and to maximise it analytically. This
is carried out in the next section.

4.3   AN EXPLICIT EXPRESSION FOR $|\bar{M}|$

A major difficulty in deriving an explicit expression for $|\bar{M}|$
from (2.1) is that the h vector consists of two types of term (see 3.3.2)

corresponding to the A and B parameters respectively. Considerable simplification occurs if either set of parameters is assumed known a priori, i.e.

  either (a) $(a_1, \ldots, a_n)$ is known; $p = m+1$

     or (b) $(b_0, \ldots, b_m)$ is known; $p = n$

Then

$$\bar{M}_{ik} = \text{Re} \int_0^\pi \delta(\omega) e^{j(i-k)\omega} d\xi(\omega), \quad i, k = 1, \ldots, p \qquad (1)$$

where

  either (a) $\delta(\omega) = |C(e^{j\omega})/DA(e^{j\omega})|^2$

     or (b) $\delta(\omega) = |CB(e^{j\omega})/DA^2(e^{j\omega})|^2$ $\qquad (2)$

Remark 1

Case (a) includes the system $A(\cdot) \equiv 1$, i.e. a weighting function model with coloured noise.

The main results are brought together in Theorem 1 below. The following lemma is needed.

Lemma 1

Let

$$t_{2k} = e^{-j\omega_k}$$
$$\qquad\qquad k = 1, 2, \ldots, \ell \qquad (3)$$
$$t_{2k-1} = e^{j\omega_k}$$

Then

$$\prod_{1 \leq i < k \leq 2\ell} (t_k - t_i) = (-2j)^\ell 2^{\ell(\ell-1)} \prod_{k=1}^{\ell} \sin \omega_k \prod_{1 \leq i < s \leq \ell} (\cos \omega_i - \cos \omega_s) \qquad (4)$$

Proof

Denote the expression on the left hand side of (4) by $\Delta_{2\ell}$. Then

$$\Delta_{2\ell} = \Delta_{2\ell-2} \prod_{1 \le i < 2\ell-1} (t_{2\ell-1} - t_i) \prod_{1 \le i < 2\ell} (t_{2\ell} - t_i)$$

$$= \Delta_{2\ell-2} (e^{-j\omega_\ell} - e^{j\omega_\ell}) \prod_{i=1}^{\ell-1} (e^{j\omega_\ell} - e^{j\omega_i}) (e^{j\omega_\ell} - e^{-j\omega_i})$$

$$\times (e^{-j\omega_\ell} - e^{j\omega_i}) (e^{-j\omega_\ell} - e^{-j\omega_i})$$

$$= -2j\Delta_{2\ell-2} \sin \omega_\ell \prod_{i=1}^{\ell-1} 4(\cos \omega_i - \cos \omega_\ell)^2$$

Now,

$$\Delta_2 = -2j \sin \omega_1$$

and the result follows.   #

* Theorem 1

Consider the process model (2.2.6) in which either A or B is known a priori and the number of unknown system parameters is 2ℓ. Then the design $\{\lambda_i, \omega_i, i=1,\ldots,\ell\}$ yields the cost

$$|\bar{M}| = \{2^{\ell(\ell-1)} \prod_{i=1}^{\ell} \lambda_i \delta(\omega_i) \sin \omega_i \prod_{1 \le s < t \le \ell} (\cos \omega_s - \cos \omega_t)^2\}^2 \qquad (5)$$

where $\delta(\omega)$ is given by (2).

If, in addition, $\delta(\omega) \equiv$ constant and the design is in $\mathcal{D}_1$ and D-optimal then

$$\lambda_i = \ell^{-1}$$

$$i = 1, \ldots, \ell \qquad (6)$$

$$\omega_i = (2i-1)\pi/2\ell$$

Proof

Using the notation of Section 2

$$|\bar{M}| = 2^{-p}\{\prod_{i=1}^{\ell} \lambda_i^2\}|F||F*| \tag{7}$$

where

$$F_{2\ell \times 2\ell} = \begin{bmatrix} \sqrt{\delta(\omega_1)}\, e^{j\omega_1} & \sqrt{\delta(\omega_1)}\, e^{-j\omega_1} & \cdots & \sqrt{\delta(\omega_\ell)}\, e^{-j\omega_\ell} \\ \sqrt{\delta(\omega_1)}\, e^{2j\omega_1} & \sqrt{\delta(\omega_1)}\, e^{-2j\omega_1} & \cdots & \sqrt{\delta(\omega_\ell)}\, e^{-2j\omega_\ell} \\ \vdots & \vdots & & \vdots \\ \sqrt{\delta(\omega_1)}\, e^{2\ell j\omega_1} & \sqrt{\delta(\omega_1)}\, e^{-2\ell j\omega_1} & \cdots & \sqrt{\delta(\omega_\ell)}\, e^{-2\ell j\omega_\ell} \end{bmatrix}$$

Then

$$|F| = \{\prod_{i=1}^{\ell} \delta(\omega_i)\} \begin{vmatrix} 1 & 1 & \cdots & 1 \\ e^{j\omega_1} & e^{-j\omega_1} & \cdots & e^{-j\omega_\ell} \\ \vdots & \vdots & & \vdots \\ e^{j\omega_1(2\ell-1)} & e^{-j\omega_1(2\ell-1)} & \cdots & e^{-j\omega_\ell(2\ell-1)} \end{vmatrix}$$

Using a property of Vandermonde matrices

$$|F| = \prod_{i=1}^{\ell} \delta(\omega_i) \prod_{1 \le i < k \le 2\ell} (t_k - t_i) \tag{8}$$

where $(t_1, \ldots, t_{2\ell})$ is given by (3).

The determinant of F* is the complex conjugate of $|F|$. Then (5) follows from (4), (7) and (8).

If the design is in $\mathcal{D}_1$ and is D-optimal, then $\lambda_i = \ell^{-1}$, $i = 1, \ldots, \ell$ by Theorem (2.1). If, in addition, $\delta(\omega) \equiv$ constant then, from (5)

$$|\bar{M}| \propto \{ \prod_{i=1}^{\ell} \sin \omega_i \prod_{1 \leq s < t \leq \ell} (\cos \omega_s - \cos \omega_t)^2 \}^2$$

$$= \prod_{i=1}^{\ell} (1-x_i^2) \prod_{1 \leq s < t \leq \ell} (x_s - x_t)^4 \quad \text{where } x_i \triangleq \cos \omega_i, \quad i = 1, \ldots, \ell$$

and

$$\frac{\partial}{\partial x_i} \log |\bar{M}| = 4 \sum_{\substack{k=1 \\ k \neq i}}^{\ell} \frac{1}{x_i - x_k} - \frac{2x_i}{1 - x_i^2} \quad i = 1, \ldots, \ell \qquad (9)$$

The minima of $|\bar{M}|$ occur if $\omega_i = 0$ or π for some i or if two frequencies are equal. The maxima of $|\bar{M}|$ occur where the right hand side of (9) vanishes, $x_i = x_i^0$, $i = 1, \ldots, \ell$ say.

Let

$$\beta(x) = \prod_{i=1}^{\ell} (x - x_i^0) \qquad (10)$$

Then

$$\frac{\beta''(x_i^0)}{\beta'(x_i^0)} = 2 \sum_{\substack{k=1 \\ k \neq i}}^{\ell} \frac{1}{x_i^0 - x_k^0} \quad \text{where } \beta'(x) \text{ denotes } d\beta/dx, \text{ etc.}$$

and the maxima of $|\bar{M}|$ occur when

$$\gamma(x_i^0) \equiv (1 - x_i^{02}) \beta''(x_i^0) - x_i \beta'(x_i^0) = 0, \quad i = 1, \ldots, \ell \qquad (11)$$

However, both $\beta(x)$ and $\gamma(x)$ are ℓ^{th} order polynomials with the same ℓ zeros. Therefore they are proportional and the constant of proportionality can be determined by equating the coefficients of x^{ℓ}. This leads to the differential equation

$$(1-x^2) \beta''(x) - x\beta'(x) + \ell^2 \beta(x) = 0 \qquad (12)$$

Writing $\chi(\omega) = \beta(\cos \omega)$, equation (12) can be cast in the form

$$\chi''(\omega) + \ell^2\chi(\omega) = 0 \tag{13}$$

where

$$\chi'(0) = -(1-x^2)^{\frac{1}{2}}\beta'(x)\big|_{x=1} = 0$$

and therefore

$$\chi(\omega) \propto \cos \ell\omega \tag{14}$$

The maxima of $|\bar{M}|$ therefore coincide with the zeros of $\cos \ell\omega$ lying in $(0,\pi)$ and this leads to the spectrum given in (6).   q.e.d. #

The weighting function model (3.5.6), i.e.

$$y_k = z^{-s}B(z^{-1})u_k + e_k \tag{15}$$

satisfies the hyperplane condition (3.5.3). If m is odd then p = m+1 is even and there exists a unique p/2-frequency (principal) D-optimal design in $\mathcal{D}_1$. Further, $\delta(\omega) \equiv 1$ and the design is given by (6).

From (1) and (6) the information matrix corresponding to the D-optimal design is given by

$$\bar{M}_{ik} = \frac{1}{\ell} \text{Re} \sum_{t=1}^{\ell} \exp\{j(i-k)(2t-1)\pi/2\ell\}, \quad i, k = 1, \ldots, m+1 = 2\ell$$

which yields

$$\bar{M} = I_{m+1}, \text{ the } (m+1)\times(m+1) \text{ unit matrix} \tag{16}$$

However, for a white noise input, i.e.

$$d\xi(\omega) = \pi^{-1}d\omega \quad \omega \in [0,\pi]$$

then from (1)

$$\bar{M}_{ik} = \pi^{-1} \operatorname{Re} \int_{0}^{\pi} e^{j(i-k)\omega} d\omega$$

leading again to (16), i.e. D-optimality.

Levin [L2] demonstrates the optimal properties of white noise inputs for weighting function models of the form (15). The above analysis shows that the deterministic input whose spectral properties are defined by (6) is also D-optimal.

The more general model

$$y_k = z^{-s} B(z^{-1}) u_k + D(z^{-1}) e_k \tag{17}$$

where m is odd and $m \geq r$ also exhibits a unique p/2-frequency D-optimal design in $\mathcal{D}_1$. However, in this case $\delta(\omega) = |D(e^{j\omega})|^{-2}$ and (9) is replaced by

$$\frac{\partial}{\partial x_i} \log |\bar{M}| = 4 \sum_{\substack{k=1 \\ k \neq i}}^{\ell} \frac{1}{x_i - x_k} - \frac{2x_i}{1-x_i^2} - 2 \frac{\varepsilon'(x_i)}{\varepsilon(x_i)}, \quad i = 1, \ldots, \ell \tag{18}$$

where $\varepsilon(x)$ is the rth order polynomial given by

$$\varepsilon(\cos \omega) = |D(e^{j\omega})|^2 \tag{19}$$

Carrying through the analysis as before leads to the differential equation

$$(1-x^2)\varepsilon(x)\beta''(x) - x\varepsilon(x)\beta'(x) - (1-x^2)\varepsilon'(x)\beta'(x) = q(x)\beta(x) \tag{20}$$

where $\beta(x)$ is given by (10) and $q(x)$ is an unknown r^{th} order polynomial.

The right hand side of (20) gives rise to nonlinear terms in the set of $\ell+r$ simultaneous equations for the coefficients of $q(x)$ and $\beta(x)$ and no analytic expression for the zeros of $\beta(x)$, corresponding to the D-optimal design spectrum, appears to be attainable.

The results for continuous-time systems corresponding to those above can be derived in a similar way.

Remark 2

A major aim of the analysis presented in both Chapter 3 and the present chapter is a reduction in the magnitude of the optimization problem. However, even with such reductions, the use of gradient methods to minimize the chosen cost function can lead to the problem of local minima and it may be necessary to try several starting designs to ensure global optimality.

Global optimality is achieved asymptotically by several sequential design procedures [W2] [F1] [A6] introduced into the static experiment design field since 1970. These are all based on the Kiefer-Wolfowitz theory [F1], the relevant points of which are developed in the next section.

4.4   THE KIEFER-WOLFOWITZ THEORY

The cost criteria introduced in Section (2.9) are usually termed parameter space criteria [e.g. M2] as their use is aimed at minimizing, in some sense, cov $\hat{\theta}$, i.e. the covariance of the parameter estimator. However, if the purpose of the identification is to accurately predict the output sequence $\{y_k, \ k=1,\ldots,N\}$ for a range of input signals, it is more reasonable to cost the variance of $y_k(\hat{\theta})$, the output from the process model using the estimated system transfer function. This leads

to the class of output space criteria [F1] [M2]. The point of
departure here for the Kiefer-Wolfowitz theory is the equivalence
of D-optimality and a certain output space criterion discussed below.

Consider the process model (2.2.4) expressed in the form

$$y_k(\hat{\theta}) = z(z^{-1}, \hat{\theta}) u_k + H(z^{-1}) e_k \tag{1}$$

$$e_k \sim N(0,1), \quad k = 1, \ldots, N$$

where dependence on the estimated system parameters is shown
explicitly. It is assumed that the actual process is described
by (1) with $\hat{\theta}$ replaced by θ, the true parameter vector (see Section 2.3).
Then, to the first order in $\hat{\theta} - \theta$,

$$y_k(\hat{\theta}) \simeq y_k(\theta) + (\hat{\theta} - \theta)^T z_\theta(z^{-1}) u_k \tag{2}$$

where

$$z_\theta(z^{-1}) = \partial z(z^{-1}, \theta) / \partial \theta$$

From (2), the variance of $y_k(\hat{\theta})$ is given by

$$\text{var } y_k(\hat{\theta}) = [z_\theta^T(z^{-1}) u_k] \text{ cov } \hat{\theta} [z_\theta(z^{-1}) u_k] \tag{3}$$

Consider N large and $\dot{u}_1, \ldots, u_N$ corresponding to a unit power sine
wave of frequency ω_0. Then (cf. Sections (2.4) and (2.6))

$$\text{var } y_k(\hat{\theta}) \simeq N z_\theta^T(e^{-j\omega_0}) \bar{M}^{-1}(\xi) z(e^{j\omega_0})$$

where ξ is the input design yielding $\hat{\theta}$. In the limit

$$\lim_{N \to \infty} \frac{1}{N} \text{var } y_k(\hat{\theta}) = |H(e^{j\omega_0})|^2 h^*(e^{j\omega_0}) \bar{M}^{-1}(\xi) h(e^{j\omega_0}) \quad (4)$$

$$= S(\omega_0) \text{ tr } [\bar{M}^{-1}(\xi) \bar{M}(\xi_{\omega_0})] \quad (5)$$

where ξ_{ω_0} denotes the single frequency design,

$$S(\omega) = |H(e^{j\omega})|^2 = |D(e^{j\omega})/C(e^{j\omega})|^2$$

denotes the spectral density of the process noise and the information matrix $\bar{M}(\xi)$ is given by

$$\bar{M}(\xi) = \text{Re} \int_0^{\pi} h(e^{j\omega}) h^*(e^{j\omega}) d\xi(\omega) \quad (6)$$

in the usual notation.

Note that ω_0 and ξ on which var $y_k(\hat{\theta})$ depends can be chosen independently.

Definition 1

The *generalized variance* $d(\omega,\xi)$ is defined as

$$d(\omega,\xi) = \lim_{N \to \infty} \frac{1}{N} \text{var } [H^{-1}(z^{-1}) y_k(\hat{\theta})] \quad (7)$$

$$= h^*(e^{j\omega}) \bar{M}^{-1}(\xi) h(e^{j\omega}) \quad (8)$$

$$= \text{tr } [\bar{M}^{-1}(\xi) \bar{M}(\xi_{\omega})], \text{ using } (6) \quad (9)$$

Note that in (7) the estimated output is passed through the noise-whitening filter. This definition of $d(\omega,\xi)$ differs from that of Mehra [M2] but leads to some notational savings. Mehra defines $d(\omega,\xi)$ as the left hand side of (4). The two definitions coincide for white process noise.

Definition 2

The design $\xi^* \in \Xi$ is said to be G-optimal if ξ^* minimises

$\max_{\omega \in [0,\pi]} d(\omega,\xi)$ in Ξ,

The central theorem, due to Mehra [M3], can now be stated.

Theorem 1 (Equivalence Theorem)

The following statements are equivalent:

(i) The normalized design ξ^* is D-optimal;

(ii) ξ^* is G-optimal;

(iii) $\max_{\omega \in [0,\pi]} d(\omega,\xi^*) = p$

Proof

For a full proof in the MIMO case, the reader is referred to Mehra [M3]. Here only parts of the proof are presented. It is shown that (ii) and (iii) follow from (i).

From (9)

$$\int_0^\pi d(\omega,\xi)\,d\xi(\omega) = \operatorname{tr}\,[\bar{M}^{-1}(\xi)\int_0^\pi \bar{M}(\xi_\omega)\,d\xi(\omega)]$$

$$= \operatorname{tr}\,I_p = p \qquad (10)$$

Therefore

$$\max_{\omega \in [0,\pi]} d(\omega,\xi) \geq p \quad \text{for all } \xi \in \Xi_1 \qquad (11)$$

Let $\xi^*,\ \xi^0 \in \Xi_1$ and $\xi = (1-\alpha)\xi^* + \alpha\xi^0 \in \Xi_1$ for some $\alpha \in [0,1]$.

Then

$$\bar{M}(\xi) = (1-\alpha)\bar{M}(\xi^*) + \alpha\bar{M}(\xi^0)$$ (12)

and

$$\frac{\partial}{\partial\alpha} \log |\bar{M}(\xi)| = \operatorname{tr} [\frac{\partial\bar{M}}{\partial\alpha} \frac{\partial}{\partial\bar{M}} \log |\bar{M}(\xi)|]$$

$$= \operatorname{tr} [(\bar{M}(\xi^0) - \bar{M}(\xi^*))\bar{M}^{-1}(\xi)]$$ (13)

where $\bar{M}(\xi)$ is assumed nonsingular.

If $\xi^0 = \xi_\omega$, then

$$[\frac{\partial}{\partial\alpha} \log |\bar{M}(\xi)|]_{\alpha=0} = d(\omega, \xi^*) - p$$ (14)

If ξ^* is D-optimal, the lefthand side of (14) cannot be positive.

Therefore

$$d(\omega, \xi^*) \leq p \quad \text{for all } \omega \in [0, \pi]$$ (15)

From (15)

$$\max_{\omega \in [0, \pi]} d(\omega, \xi^*) \leq p$$

and from (11)

$$\max_{\omega \in [0, \pi]} d(\omega, \xi^*) \geq p$$

Therefore

$$\max_{\omega \in [0, \pi]} d(\omega, \xi^*) = p$$ (16)

as required, i.e. (i) implies (iii).

From (11) and (16) it follows that ξ^* minimizes $\max_{\omega \in [0, \pi]} d(\omega, \xi)$ in Ξ_1,

i.e. (i) implies (ii).   #

Corollary 1

If ω_i belongs to the spectrum of ξ^* then

$$d(\omega_i, \xi^*) = p \tag{17}$$

Proof

Assume $d(\omega_i, \xi^*) < p$ for some ω_i in the spectrum of ξ^*. Then, using (16)

$$\int_0^\pi d(\omega, \xi^*) \, d\xi^*(\omega) < p$$

contradicting (10).   #

Remark 1

The corollary states a necessary (not sufficient) condition for D-optimality and (17) provides a useful test for possible D-optimal designs.

Remark 2

The Equivalence Theorem can be extended to more general cost criteria [K6] [M3] but these generalizations are not considered here.

The generation of D-optimal input designs using sequential procedures is discussed in the next section.

4.5   A SEQUENTIAL DESIGN PROCEDURE

A sequential design algorithm with proven convergence to a D-optimal design was proposed in 1970 by Wynn [W2] for static experimental design. A number of authors [F1] [A6] [S4] have proposed variations in order to improve the rate of convergence. The following algorithm is essentially due to Fedorov [F1] but placed in the context of dynamic system input design by Mehra [M3].

Algorithm 1

(i) Choose a design $\xi_0 \in \Xi_1$ such that $I(\xi_0) \geq p/2$

(ii) Set $k = 0$

(iii) Choose ω_k s.t. $d(\omega_k, \xi_k) = \max\limits_{\omega \in [0,\pi]} d(\omega, \xi_k)$

(iv) If $d(\omega_k, \xi_k) = p$, stop

(v) Update design to $\xi_{k+1} = (1-\alpha_k)\xi_k + \alpha_k \xi_{\omega_k} \in \Xi_1$

(vi) Set $k = k+1$; go to (iii)

Theorem 1

If the sequence $\{\alpha_k\}$ in Algorithm 1 is chosen so that

$$\alpha_k \in [0,1], \lim_{k \to \infty} \alpha_k = 0, \sum_{k=0}^{\infty} \alpha_k = \infty \tag{1}$$

then

$$\lim_{k \to \infty} \xi_k = \xi*$$

a D-optimal design in Ξ_1.

Proof

To the first order in α_k

$$\log |\bar{M}(\xi_{k+1})| \simeq \log |\bar{M}(\xi_k)| + \alpha_k \{d(\omega_k, \xi_k) - p\} \tag{2}$$

From (1), $\lim\limits_{k \to \infty} \alpha_k = 0$ and therefore $\exists\ k_0$, η dependent on k_0 s.t. $0 < \eta \leq 1$ and $\forall\ k \geq k_0$,

$$\log |\bar{M}(\xi_{k+1})| > \log |\bar{M}(\xi_k)| + \eta\alpha_k \{d(\omega_k, \xi_k) - p\} \tag{3}$$

Step (iii) of the algorithm ensures that $d(\omega_k, \xi_k) \geq p$, $\forall\ k \geq 0$ and therefore

$$|\bar{M}(\xi_{k_0})| < |\bar{M}(\xi_{k_0+1})| < \ldots < |\bar{M}(\xi^*)| \tag{4}$$

i.e. $\{|\bar{M}(\xi_k)|, k \geq k_0\}$ is a monotonically increasing sequence bounded above. Therefore the sequence converges, i.e.

$$\lim_{k \to \infty} |\bar{M}(\xi_k)| = |\bar{M}(\xi')| \leq |\bar{M}(\xi^*)|$$

for some $\xi' \in \Xi_1$.

Assume that $|\bar{M}(\xi')| \neq |\bar{M}(\xi^*)|$. Then $\exists \; \varepsilon \; \text{s.t.} \; d(\omega_k, \xi_k) - p \geq \varepsilon > 0$, $\forall \; k \geq 0$, and from (3)

$$\log |\bar{M}(\xi')| > \log |\bar{M}(\xi_{k_0})| + \eta\varepsilon \sum_{k=k_0}^{\infty} \alpha_k$$

Therefore, if $\sum_{k=0}^{\infty} \alpha_k = \infty$ the sequence $\{|\bar{M}(\xi_k)|\}$ is unbounded. This contradicts (4) and it follows that $|\bar{M}(\xi')| = |\bar{M}(\xi^*)|$ as required. q.e.d. #

Remark 1

In general, $d(\omega, \xi_k)$ has a number of stationary points and step (iii) must be carried out by a grid search. Time can be saved by choosing ω_k as the smallest value of ω for which $d(\omega_k, \xi_k) > p$. The existence of such an ω_k is guaranteed by (4.11) if $\xi_k \neq \xi^*$ and the convergence proof goes through as above. However, the rate of convergence of the algorithm may be considerably worsened.

Remark 2

The stopping condition (iv) is a gradient criterion and can be replaced by a test of the form

$$|\det \bar{M}(\xi_{k+1}) - \det \bar{M}(\xi_k)| < \varepsilon \det \bar{M}(\xi_k)$$

for some $\varepsilon > 0$.

The updating of $\bar{M}^{-1}(\xi_k)$ and $|\bar{M}(\xi_k)|$ from loop to loop is simplified by the following well known result:

Lemma 1

If M is a nonsingular pxp matrix and f, g are two p vectors then the pxp matrix M_1 defined by

$$M_1 = M + fg^T \tag{5}$$

has determinant

$$|M_1| = |M|(1+g^T M^{-1}f) \tag{6}$$

and inverse

$$M_1^{-1} = \{I - \frac{M^{-1}fg^T}{1+g^T M^{-1}f}\}M^{-1} \tag{7}$$

Step (v) of Algorithm (5.1) leads to

$$\bar{M}(\xi_{k+1}) = (1-\alpha_k)\bar{M}(\xi_k) + \alpha_k \text{ Re } h(e^{j\omega_k})h^*(e^{j\omega_k})$$

$$= (1-\alpha_k)\bar{M}(\xi_k) + \tfrac{1}{2}\alpha_k h(e^{j\omega_k})h^*(e^{j\omega_k}) + \tfrac{1}{2}\alpha_k \bar{h}(e^{j\omega_k})h^T(e^{j\omega_k})$$

Applying the formula of Lemma 1 twice, after some messy manipulation

$$\frac{|\bar{M}(\xi_{k+1})|}{|\bar{M}(\xi_k)|} = (1-\alpha_k)^P\{1 + \frac{\alpha_k}{1-\alpha_k} d(\omega_k,\xi_k) + (\frac{\alpha_k}{1-\alpha_k})^2 g(\omega_k,\xi_k)\} \tag{8}$$

where

$$g(\omega,\xi) = \tfrac{1}{4}\{d^2(\omega,\xi) - |d_1(\omega,\xi)|^2\} \tag{9}$$

and

$$d_1(\omega,\xi) = h^T(e^{j\omega})\bar{M}^{-1}(\xi)h(e^{j\omega}) \tag{10}$$

Also

$$\bar{M}^{-1}(\xi_{k+1}) = (1+2\gamma_k)\bar{M}^{-1}(\xi_k) + \frac{2\gamma_k}{(1+2\gamma_k)^{p-1}} \frac{|\bar{M}(\xi_k)|}{|\bar{M}(\xi_{k+1})|} \times$$

$$x \ \text{Re} \ [\bar{M}^{-1}(\xi_k)h(e^{j\omega_k})\{\gamma_k\bar{d}_1(\omega_k,\xi_k) \ \bar{M}^{-1}(\xi_k)h(e^{j\omega_k})]^T$$

$$- (1+\gamma_k d(\omega_k,\xi_k))[\bar{M}^{-1}(\xi_k)h(e^{j\omega_k})]*\}] \qquad (11)$$

where

$$2\gamma_k = \alpha_k/(1-\alpha_k)$$

and

$$d(\omega_k,\xi_{k+1}) = [\frac{1+2\gamma_k}{1+\gamma_k d(\omega_k,\xi_k)}] \ [d(\omega_k,\xi_k) -$$

$$- \frac{|d_1(\omega_k,\xi_k)|^2}{[\{1+\gamma_k d(\omega_k,\xi_k)\}^2 - \gamma_k^2|d_1(\omega_k,\xi_k)|^2]}] \qquad (12)$$

If $\alpha_k = 1$, a relevant possibility if $p \leq 2$, then clearly expressions (8) - (12) can be used only after clearing any factors $(1-\alpha_k)$ from denominators.

Remark 3

Let $\Omega_d(\xi) = \{\omega|d(\omega,\xi) = \max_{\omega' \in [0,\pi]} \ d(\omega',\xi)\}$. If $\Omega_d(\xi_k)$ has more

than one element, then it is reasonable to choose

$\omega_k = \arg\max_{\omega \in \Omega_d(\xi_k)} \ g(\omega,\xi_k) = \arg\min_{\omega \in \Omega_d(\xi_k)} \ |d_1(\omega,\xi_k)|$. The convergence

proof remains unchanged. #

The quantities $d(\omega,\xi)$ and $g(\omega,\xi)$ defined in (4.8) and (9) respectively

have the properties

$$d(\omega,\xi) > 0$$

$$\forall\ \omega\ \epsilon\ [0,\pi] \qquad\qquad (13)$$

$$0 \le g(\omega,\xi) \le d^2(\omega,\xi)/4$$

Further:

* **Result 1**

The quantity $g(\omega,\xi)$ is zero iff $\omega = 0$ or π.

Proof

Let $h_1(\omega)$, $h_2(\omega)$ denote the real and imaginary parts respectively of $h(e^{j\omega})$ and define

$$<u,v> \underline{\Delta}\ u^T \overline{M}^{-1}(\xi)v; \quad ||u|| = <u,u>^{\frac{1}{2}}$$

for arbitrary real p-vectors u, v.

Then

$$d(\omega,\xi) = <h_1(\omega),h_1(\omega)> + <h_2(\omega),h_2(\omega)> \qquad (14)$$

and

$$g(\omega,\xi) = <h_1(\omega),h_1(\omega)><h_2(\omega),h_2(\omega)> - <h_1(\omega),h_2(\omega)>^2$$

$$= ||h_1(\omega) - \frac{<h_1(\omega),h_2(\omega)>}{<h_2(\omega),h_2(\omega)>} h_2(\omega)||^2 \qquad (15)$$

so that

$$g(\omega,\xi) = 0 \Leftrightarrow h_1(\omega),h_2(\omega) \qquad \text{are linearly dependent}$$

$$\Leftrightarrow h(e^{j\omega}),\bar{h}(e^{j\omega}) \qquad \text{are linearly dependent}$$

$$\Leftrightarrow \omega = 0 \text{ or } \pi \qquad \text{(see Theorem 2.8.2)}$$

This is the required result. #

In the next section, equation (12) is used to derive some further properties of D-optimal designs.

4.6   FURTHER PROPERTIES OF D-OPTIMAL DESIGNS

The design measures ξ_k, $k = 0$, 1, ... generated sequentially by Algorithm (5.1) belong to $\mathcal{D}_1$ or Ξ_1 iff the initial design ξ_0 belongs to $\mathcal{D}_1$ or Ξ_1 respectively.

In practice starting designs are selected from $\mathcal{D}_1$; however, for completeness it is assumed here that $\xi_0 \in \Xi_1$ and may have a mixed spectrum.   Then the following definition is of use.

Definition 1

Define

$$\lambda(\xi,\omega) = \xi(\omega) - \xi(\omega-), \quad \forall \xi \in \Xi_1, \ \omega \in [0,\pi]$$

where

$$\xi(\pi) = 1, \quad \xi(0-) = 0$$

Then

$$\lambda(\xi,\omega) = \lambda(\xi_d,\omega), \quad \forall \omega \in [0,\pi]$$

where ξ_d is the component of ξ with discrete spectrum.   If ξ_d has the representation $\{\lambda_i,\omega_i, i=1,\ldots,\ell\}$ then

$$\lambda(\xi,\omega) = \lambda_k \text{ if } \omega = \omega_k, \quad k = 1, \ldots, \ell$$

$$= 0 \quad \text{otherwise}$$

If $\xi_d(\omega) \equiv 0$, then $\lambda(\xi,\omega) \equiv 0$.

The following theorem is an extension of a result due to Atwood [A6].

* **Theorem 1**

If $\xi^* \in \Xi_1$ is D-optimal then

$$\lambda(\xi^*,\omega) \leq \frac{2}{p}, \quad \forall \omega \in [0,\pi] \tag{1}$$

Proof

The result is trivial for $p = 1$, 2 and Theorem (2.1) covers the case $I(\xi^*) = p/2$. It is therefore assumed that $p \geq 3$ and $I(\xi^*) > p/2$. This ensures that ξ^* is at least a two-frequency design and therefore

$$\lambda(\xi^*,\omega) < 1, \quad \forall \omega \in [0,\pi] \tag{2}$$

Consider the design measure $\xi \in \Xi_1$ defined by

$$\xi(\omega) = (1-\alpha_1)\xi^*(\omega) + \alpha_1 \xi_{\omega'}(\omega) \tag{3}$$

where $\xi_{\omega'}$ is the single frequency ω' design measure and

$$\alpha_1 = \lambda(\xi^*,\omega')[\lambda(\xi^*,\omega')-1]^{-1} < 0$$

or, defining

$$\alpha = \lambda(\xi^*,\omega') \tag{4}$$

and rearranging (3)

$$\xi^*(\omega) = (1-\alpha)\xi(\omega) + \alpha\xi_{\omega'}(\omega) \tag{5}$$

The choice of α in (4) ensures that $\lambda(\xi,\omega') = 0$. Of course, $\lambda(\xi^*,\omega')$

may be zero.

From (5.12) and (5.13)

$$d(\omega',\xi^*) \leq \left[\frac{(2-\alpha)d(\omega',\xi)}{2(1-\alpha)+\alpha d(\omega',\xi)}\right] < \frac{2}{\alpha} = \frac{2}{\lambda(\xi^*,\omega')} \qquad (6)$$

From Corollary (4.1), $d(\omega',\xi^*) = p$. Therefore $\lambda(\xi^*,\omega') < 2/p$

where ω' is arbitrary in $[0,\pi]$. The equality in (1) occurs for

$I(\xi^*) = p/2$ (see Theorem 2.1). This completes the proof. #

Remark 1

If $\xi^* \in \mathcal{D}_1$ then, in the usual notation, $\lambda_k \leq 2/p$, $k = 1, \ldots, \ell$. #

* Theorem 2

If there exists a principal D-optimal design ξ^* in $\mathcal{D}_1$ with

representation $\{\lambda_i, \omega_i, i=1,2,\ldots\ell\}$ then

$$d_1(\omega_i,\xi^*) = 0 \qquad \text{if } \omega_i \in (0,\pi)$$

$$i = 1, 2, \ldots, \ell \quad (7)$$

$$|d_1(\omega_i,\xi^*)| = p \quad \text{if } \omega_i = 0 \text{ or } \pi$$

Proof

The result for $\omega_i = 0$ or π follows immediately from Corollary

(4.1) and result (5.1).

Consider ξ_1, $\xi_{\omega_0} \in \mathcal{D}_1$ and define ξ by

$$\xi = (1-\alpha)\xi_1 + \alpha\xi_{\omega_0} \qquad (8)$$

Then ξ is a valid design measure in $\mathcal{D}_1$ provided $\alpha \in [-\lambda(\xi_1,\omega_0)\{1-\lambda(\xi_1,\omega_0)\}^{-1}, 1]$.

Note that a negative value of α is allowed if $\lambda(\xi_1,\omega_0) > 0$ and corresponds

to removal of power from a particular frequency in the spectrum of ξ_1.

From (5.8)

$$\left|\bar{M}(\xi)\right| = \left|\bar{M}(\xi_1)\right| (1+\beta)^{-p}\{1+\beta d(\omega_0,\xi_1) + \beta^2 g(\omega_0,\xi_1)\} \qquad (9)$$

where

$$\beta = \alpha/(1-\alpha) \in [-\lambda(\xi_1,\omega_0),\infty]$$

Let $\xi_1 = \xi^*$, $\omega_0 = \omega_i$ and $\beta = -\lambda(\xi^*,\omega_i) = -\lambda_i$.

This choice of β corresponds to the complete removal of ω_i from the spectrum of ξ^*. Therefore $I(\xi) < p/2$ and $\left|\bar{M}(\xi)\right| = 0$.

Therefore

$$1 - \lambda_i d(\omega_i,\xi^*) + \lambda_i^2 g(\omega_i,\xi^*) = 0 \qquad (10)$$

Using Corollary (4.1) and substituting for $g(\omega,\xi)$ from (5.9) into (10) yields

$$(\lambda_i p-2)^2 = \lambda_i^2 \left|d_1(\omega_i,\xi^*)\right|^2 \qquad (11)$$

From Theorem (2.1), $\lambda_i = 2/p$ or $1/p$ and therefore the positive square root must be taken in (11), i.e.

$$\left|d_1(\omega_i,\xi^*)\right| = p - 2/\lambda_i \qquad (12)$$

The result then follows from Theorem (2.1)   #

It also follows that

$$g(\omega_i,\xi^*) = p^2/4 \text{ if } \omega_i \in (0,\pi) \qquad (13)$$

$$= 0 \qquad \text{if } \omega_i = 0 \text{ or } \pi \qquad (14)$$

but note that (14) is a particular case of result (4.1).

Example 1

Consider the model discussed in Section 3, i.e.

$$y_k = z^{-s} B(z^{-1}) u_k + e_k \tag{15}$$

where $p = m+1$ is even. Then the lower principal D-optimal design in $\mathcal{D}_1$ is given by

$$\lambda_i = \ell^{-1}$$
$$i = 1, \ldots, \ell \tag{16}$$
$$\omega_i = (2i-1)\pi/2\ell$$

where $2\ell = m+1$. This yields $\bar{M}(\xi^*) = I_p$.

From (15)

$$h_i(e^{j\omega}) = e^{ij\omega} \qquad i = 0, 1, \ldots, m \tag{17}$$

and therefore

$$d(\omega_i, \xi^*) = [1 \quad e^{-j\omega_i} \ldots e^{-mj\omega_i}] \begin{bmatrix} 1 \\ e^{j\omega_i} \\ \vdots \\ e^{mj\omega_i} \end{bmatrix} = m+1 = p \tag{18}$$

and

$$d_1(\omega_i, \xi^*) = \sum_{k=0}^{m} e^{2kj\omega_i} = \frac{1 - e^{2j(m+1)\omega_i}}{1 - e^{2j\omega_i}}$$

$$= \frac{1 - e^{4j\ell\omega_i}}{1 - e^{2j\omega_i}} = 0 \qquad \text{as required.}$$

4.7   A FURTHER ALGORITHM FOR SEQUENTIAL DESIGN

The remainder of theis chapter is concerned with possible extensions of the sequential design algorithm discussed in Section 5, i.e.

Algorithm 1

$$\xi_{k+1} = (1-\alpha_k)\xi_k + \alpha_k \xi_{\omega_k} \tag{1}$$

where

$$d(\omega_k, \xi_k) = \max_{\omega \in [0,\pi]} d(\omega, \xi_k) \tag{2}$$

and

$$\alpha_k \in [0,1], \lim_{k \to \infty} \alpha_k = 0, \sum_{k=0}^{\infty} \alpha_k = \infty \tag{3}$$

Then

$$|\bar{M}(\xi_{k+1})| = |\bar{M}(\xi_k)| \phi(\omega_k, \beta_k, \xi_k) \tag{4}$$

where

$$\beta_k = \alpha_k/(1-\alpha_k) \tag{5}$$

and

$$\phi(\omega, \beta, \xi) \triangleq (1+\beta)^{-p}\{1 + \beta d(\omega, \xi) + \beta^2 g(\omega, \xi)\} \tag{6}$$

In terms of β_k, the conditions (3) can be expressed as

$$\beta_k \in [0, \infty], \lim_{k \to \infty} \beta_k = 0, \sum_{k=0}^{\infty} \beta_k = \infty \tag{7}$$

Note that in any algorithm $\alpha_k = 1$ ($\beta_k = \infty$) can only be a valid choice if p = 1 or 2. In these cases, the third condition (7) must hold with the unbounded terms removed from the sum.

The expression (6) suggests that the following algorithm may be an improvement on Algorithm 1.

Algorithm 2

Choose β_k to satisfy conditions (7) and ω_k to satisfy

$$\phi(\omega_k, \beta_k, \xi_k) = \max_{\omega \in \Omega} \phi(\omega, \beta_k, \xi_k) \tag{8}$$

where

$$\Omega = \{\omega | 0 \leq \omega \leq \pi\}$$

Then:

* Theorem 1

The sequence of designs generated by Algorithm 2 converges to a D-optimal design in Ξ_1.

Proof

If $\xi_k \neq \xi^*$, a D-optimal design, then

$$\exists \ \omega_k' \in \Omega \ \text{s.t.} \quad d(\omega_k', \xi_k) > p \quad \text{(Theorem 4.1)} \tag{9}$$

Let

$$\psi(\omega, \beta, \xi) = (1+\beta)^{-p}\{1 + \beta d(\omega, \xi)\} \tag{10}$$

Then, for small β,

$$\psi(\omega, \beta, \xi) \simeq 1 + \beta\{d(\omega, \xi) - p\}$$

i.e. $\exists \ k_0$, η dependent on k_0 s.t. $0 < \eta \leq 1$ and

$$\psi(\omega_k', \beta_k, \xi_k) \geq 1 + \eta \beta_k \{d(\omega_k', \xi_k) - p\}, \quad \forall \ k \geq k_0 \tag{11}$$

Now

$$\max_{\omega \in \Omega} \phi(\omega, \beta_k, \xi_k) \geq \max_{\omega \in \Omega} \psi(\omega, \beta_k, \xi_k) \geq \psi(\omega_k', \beta_k, \xi_k) \tag{12}$$

Therefore, from (11) and (12)

$$\frac{|\bar{M}(\xi_{k+1})|}{|\bar{M}(\xi_k)|} = \max_{\omega \in \Omega} \phi(\omega, \beta_k, \xi_k) \geq 1, \ \forall \ k \geq k_0$$

where the inequality is strict unless $\xi_k = \xi*$ for some $k \geq k_0$.

It follows that either $\xi_k = \xi*$ for some finite k or that the sequence

of determinants satisfies

$$|\bar{M}(\xi_{k_0})| < |\bar{M}(\xi_{k_0+1})| < \ldots < |\bar{M}(\xi*)| \tag{13}$$

In the former case the theorem is proved. In the latter case (13)

implies that the sequence of determinants converges to $|\bar{M}(\xi')|$,

say, where $\xi' \in \Xi_1$.

If it is assumed that $|\bar{M}(\xi')| < |\bar{M}(\xi*)|$, then the same

contradiction arises as in the proof of Theorem (5.1). Therefore

$$\lim_{k \to \infty} |\bar{M}(\xi_k)| = |\bar{M}(\xi*)| \qquad \qquad \text{q.e.d.} \quad \#$$

The choice of β_k in Algorithms 1 and 2 leads to a slow rate of

convergence in general, e.g. the sequence of determinants may not

settle down into a monotone pattern (see 13) until k_0 is large.

This particular disadvantage is eliminated in the algorithms introduced

in the following. Such improvements are possible if greater freedom

is allowed in the choice of β_k. However, at this stage it is useful

to construct a more general framework to encompass a number of

sequential procedures of interest.

4.8   AN EXTENSION TO THE EQUIVALENCE THEOREM

The convergence proofs for Algorithms 1 and 2 rely on the same

contradiction, which arises when (roughly) β_k can become arbitrarily small and ω_k can be chosen so that $d(\omega_k, \xi_k) > p$. This suggests that similar proofs for more ambitious algorithms may be possible if such choices for β_k and ω_k remain open.

Introduce the sets

$$B(\beta_0) = \{\beta \mid 0 \leq \beta \leq \beta_0, \beta_0 > 0\} \tag{1}$$

$$\Omega(\omega_0, \xi) = \{\omega \mid \omega = \omega_0 \in \Omega, \begin{array}{l} d(\omega_0, \xi) = p \text{ if } \xi \text{ is D-optimal} \\ d(\omega_0, \xi) > p \text{ otherwise} \end{array} \} \tag{2}$$

$$S(\omega_0, \beta_0, \xi) = \Omega(\omega_0, \xi) \times B(\beta_0) \tag{3}$$

where $\xi \in \Xi_1$.

Note that the single-element set $\Omega(\omega_0, \xi)$ is dependent on a particular process model through $d(\omega, \xi)$. However this is not shown explicitly as different algorithms are usually compared for a single process and no confusion can arise.

Definition 1

A compact set P of pairs (ω, β) is called an S-set for ξ if there exists a pair (ω_0, β_0) such that

$$P \supseteq S(\omega_0, \beta_0, \xi) \tag{4}$$

Remark 1

Not all S-sets are dependent on a specific design measure, e.g. $B(\beta_0) \times \Omega$ is an S-set for any member of Ξ_1. In general, ω_0 is dependent on ξ.

Remark 2

It follows from (4) that if $S_2 \supseteq S_1$ and S_1 is an S-set for ξ, then so is the set S_2. This inclusion property of S-sets is crucial for establishing a class of globally convergent sequential design algorithms.

The following result extends the Kiefer-Wolfowitz Equivalence Theorem (Theorem 4.1):

* ## Theorem 1

If $P(\xi)$ is an S-set for each ξ in Ξ_1 such that $I(\xi) \geq p/2$, then the following statements are equivalent.

    (i)   The normalized design ξ^* is D-optimal

    (ii) ξ^* minimises $\max\limits_{P(\xi)} \phi(\omega,\beta,\xi)$ in Ξ_1

    (ii) $\max\limits_{P(\xi^*)} \phi(\omega,\beta,\xi^*) = 1$

Proof

Consider the sequential design procedure in which

$$\phi(\omega_k,\beta_k,\xi_k) = \max\limits_{P(\xi_k)} \phi(\omega,\beta,\xi_k) \tag{5}$$

at the $(k+1)^{st}$ stage, i.e.

$$\frac{|\bar{M}(\xi_{k+1})|}{|\bar{M}(\xi_k)|} = \max\limits_{P(\xi_k)} \phi(\omega,\beta,\xi_k) \tag{6}$$

To prove the theorem it is sufficient to prove that (i) $\Longleftrightarrow$ (iii) and (ii) $\Longleftrightarrow$ (iii).

(i) $\Rightarrow$ (iii)

Let $\xi_0 = \xi^*$, a D-optimal design. Then from (6) it follows that

$$\frac{|\bar{M}(\xi_1)|}{|\bar{M}(\xi*)|} = \max_{P(\xi*)} \phi(\omega,\beta,\xi*) \tag{7}$$

therefore

$$\max_{P(\xi*)} \phi(\omega,\beta,\xi*) \leq 1 \tag{8}$$

However, $P(\xi*)$ is an S-set and therefore

$$P(\xi*) \supseteq S(\omega_0,\beta_0,\xi*) \supset \Omega(\omega_0,\xi*) \times \{\beta|\beta=0\} \tag{9}$$

for some (ω_0,β_0). But

$$\phi(\omega,0,\xi) = 1, \ \forall \ \omega \in \Omega, \ \xi \in \Xi_1 \tag{10}$$

Therefore (8) yields $\max\limits_{P(\xi*)} \phi(\omega,\beta,\xi*) = 1$ as required.

(iii) $\Rightarrow$ (i)

Assume that $\max\limits_{P(\xi_k)} \phi(\omega,\beta,\xi_k) = 1$ but that ξ_k is not D-optimal.

Then

$$\max_{P(\xi_k)} \phi(\omega,\beta,\xi_k) \geq \max_{S(\omega_0,\beta_0,\xi_k)} \phi(\omega,\beta,\xi_k) = \max_{B(\beta_0)} \phi(\omega_0,\beta,\xi_k)$$

and $d(\omega_0,\xi_k) > p$ so that, whatever the value of β_0, $\exists \ \varepsilon \in (0,\beta_0]$ s.t.
$\phi(\omega_0,\beta,\xi_k) > 1, \ \forall \ \beta \in (0,\varepsilon) \subset B(\beta_0)$. Therefore $\max\limits_{P(\xi_k)} \phi(\omega,\beta,\xi_k) > 1$.
Contradiction.

(ii) $\Leftrightarrow$ (iii)

The above analysis shows that

$$\max_{P(\xi)} \phi(\omega,\beta,\xi) \geq 1, \ \forall \ \xi \in \Xi_1$$

Therefore ξ' minimises $\max\limits_{P(\xi)} \phi(\omega,\beta,\xi)$ in Ξ_1 iff $\max\limits_{P(\xi')} \phi(\omega,\beta,\xi') = 1$.   q.e.d. #

The above proof already indicates how the concept of S-sets may

be useful in analysing sequential design procedures.

4.9   S-ALGORITHMS AND GLOBAL CONVERGENCE

Each sequential design algorithm is characterised by a step of

the form:

$$\phi(\omega_k,\beta_k,\xi_k) = \max\limits_{S(\xi_k)} \phi(\omega,\beta,\xi_k) \tag{1}$$

where $S(\xi_k)$ is the feasible set at the $(k+1)^{st}$ stage of the algorithm.

Denoting the feasible sets for Algorithms 1 and 2 of Section 7

by $S_1(\cdot)$, $S_2(\cdot)$ respectively, then:

Algorithm 1

$$S_1(\xi_k) = \{(\omega,\beta)\,|\,\omega=\omega_f, d(\omega_f,\xi_k) = \max\limits_{\omega'\in\Omega} d(\omega',\xi_k), \beta = k^{th}\ \text{term}$$

$$\text{of series satisfying conditions (7.7)}\} \tag{2}$$

Algorithm 2

$$S_2(\xi_k) = \{(\omega,\beta)\,|\,\omega\in\Omega,\ \beta = k^{th}\ \text{term of series satisfying}$$

$$\text{conditions (7.7)}\} \tag{3}$$

The set $S_1(\xi_k)$ contains a single element and therefore the max

operation in (1) is trivial. Remark (5.3) discusses briefly a

possible modification. The set $S_2(\xi_k)$ is independent of ξ_k.

Neither $S_1(\cdot)$ nor $S_2(\cdot)$ is an S-set. The following algorithms,
however , have feasible sets which are S-sets:

Algorithm 3

(i)   Choose ω_k s.t. $d(\omega_k,\xi_k) = \max\limits_{\omega\in\Omega} d(\omega,\xi_k)$

(ii) Choose β_k s.t. $\phi(\omega_k,\beta_k,\xi_k) = \max\limits_{\beta\in B} \phi(\omega_k,\beta,\xi_k)$ where

$B = \{\beta\,|\,0\le\beta<\infty\}$.

Denoting the feasible set for Algorithm i by $S_i(\cdot)$, then

$$S_3(\xi_k) = \{(\omega,\beta)\,|\,\omega=\omega_f, d(\omega_f,\xi_k) = \max\limits_{\omega'\in\Omega} d(\omega',\xi_k), \beta\in B\} \qquad (4)$$

and

$$S_3(\xi_k) \supseteq S(\omega_f,\beta_0,\xi_k), \quad \beta_0 \text{ arbitrary} \qquad (5)$$

Remark 1

This algorithm is due to Fedorov [F1] who proves global convergence
in the case of static systems. Mehra proposes this sequential
procedure for dynamic systems, but his proof of global convergence is
incomplete [M3, pp. 39-40].

Algorithm 4

Choose ω_k, β_k s.t. $\phi(\omega_k,\beta_k,\xi_k) = \max\limits_{\Omega\times B} \phi(\omega,\beta,\xi_k)$, i.e.

$$S_4(\xi_k) = \Omega \times B \supset S(\omega_0,\beta_0,\xi_k), \quad (\omega_0,\beta_0) \text{ arbitrary} \qquad (6)$$

Algorithm 5

Atwood [A 6] points out that the removal of power from a frequency
in the current design may lead to an improved cost value over any
possible frequency addition.

Allowing for this possibility in Algorithm 3 leads to

$$S_5(\xi_k) = S_3(\xi_k) \cup S_r(\xi_k) \supseteq S_3(\xi_k) \qquad (7)$$

where

$$S_r(\xi_k) = \{(\omega,\beta) | \omega\in\Omega, \beta\in[-\lambda(\xi_k,\omega),0)\} \qquad (8)$$

Algorithm 6

Allowing for such power removal in Algorithm 4 yields the feasible set

$$S_6(\xi_k) = \{(\omega,\beta) | \omega\in\Omega, \beta\in[-\lambda(\xi_k,\omega),\infty]\}$$

$$= S_4(\xi_k) \cup S_r(\xi_k) \supseteq S_4(\xi_k) \qquad (9)$$

The relations (5), (6), (7) and (9) imply that algorithms 3, 4, 5 and 6 have feasible sets that are S-sets for each ξ_k. This suggests the following definition:

Definition 1

An algorithm is an S-*algorithm* if the feasible set $S(\xi)$ is an S-set for each ξ in Ξ_1 satisfying $I(\xi) \geq p/2$.

The main convergence result can now be stated briefly.

* Theorem 1

Every S-algorithm generates a sequence of design measures in Ξ_1 converging to a D-optimum.

Proof

The result follows in a straightforward way from the extended equivalence theorem (8.1).

Consider the S-algorithm with feasible set $P(\xi_k)$ at the $(k+1)^{st}$ stage.

If ξ_k is not D-optimal, then (Theorem 8.1(ii))

$$\frac{|\bar{M}(\xi_{k+1})|}{|\bar{M}(\xi_k)|} = \max_{P(\xi_k)} \phi(\omega,\beta,\xi_k) > 1 \tag{10}$$

Therefore the sequence of determinants is monotonically increasing and bounded above by $|\bar{M}(\xi^*)|$, where $\xi^* \in \Xi_1$ and is D-optimal. This implies convergence of the sequence to $|\bar{M}(\xi')|$, say.

Then

$$\lim_{k\to\infty} \frac{|\bar{M}(\xi_{k+1})|}{|\bar{M}(\xi_k)|} = 1 \tag{11}$$

If $|\bar{M}(\xi_k)| = |\bar{M}(\xi^*)|$ for some finite k, the theorem is proved. Otherwise, assume that $|\bar{M}(\xi')| < |\bar{M}(\xi^*)|$. Then $\xi_k \neq \xi^*$ and (10) holds for all $k \geq 0$ so that $\exists\ \delta > 0$ s.t.

$$\frac{|\bar{M}(\xi_{k+1})|}{|\bar{M}(\xi_k)|} \geq 1 + \delta > 1,\ \forall\ k \geq 0 \tag{12}$$

But (12) contradicts (11). Therefore $|\bar{M}(\xi')| = |\bar{M}(\xi^*)|$. q.e.d. #

The Algorithms 3 - 6 are S-algorithms and global convergence is therefore guaranteed. A comparison of the behaviour of these algorithms is carried out in the Appendix.

A disadvantage of purely additive procedures, e.g. Algorithms 1 - 4, is that the final design that is accepted as approximately D-optimal may contain a large number of frequencies. This can be aggravated by a bad choice of initial design. The removal Algorithms 5 and 6 in general overcome these disadvantages and produce final designs with low index.

Another possibility is to reduce the number of frequencies by a final 'rounding-off' [F1] [S1] [G5]. Of course, this is an ad hoc procedure. In the next section a sequential round-off is proposed.

4.10 ROUNDING-OFF

Consider any sequential design algorithm of the type discussed
throughout this chapter, i.e. in which

$$\xi_{k+1} = (1-\alpha_k)\xi_k + \alpha_k \xi_{\omega_k} \tag{1}$$

If ω_k belongs to the spectrum of ξ_k then no rounding-off is
necessary. Therefore assume that ω_k does not belong to the spectrum
of ξ_k but that ω_i does and

$$\omega_i = \omega_k + \epsilon \tag{2}$$

where $|\epsilon|$ is small (in some sense to be made precise).

Let

$$\lambda_i = \lambda(\xi_k,\omega_i) \quad \text{and} \quad \lambda_k = \lambda(\xi_{k+1},\omega_k) = \alpha_k$$

and consider the effect of replacing the two frequencies ω_i, ω_k by
a single frequency ω' given by

$$\omega' = \omega_k + \epsilon', \quad |\epsilon'| \text{ small} \tag{3}$$

and with weight $\lambda_i + \lambda_k$. The new information matrix $\bar{M}'_{k+1}$ is given
by

$$\bar{M}'_{k+1} = \bar{M}(\xi_{k+1}) - \lambda_i \bar{M}(\xi_{\omega_i}) - \lambda_k \bar{M}(\xi_{\omega_k}) + (\lambda_i+\lambda_k)\bar{M}(\xi_{\omega'}) \tag{4}$$

and, to the second order in ϵ, ϵ',

$$\bar{M}(\xi_{\omega_i}) \simeq \bar{M}(\xi_{\omega_k}) + \epsilon \bar{M}_\omega(\xi_{\omega_k}) + \tfrac{1}{2}\epsilon^2 \bar{M}_{\omega\omega}(\xi_{\omega_k})$$

(5)

$$\bar{M}(\xi_{\omega'}) \simeq \bar{M}(\xi_{\omega_k}) + \epsilon' \bar{M}_\omega(\xi_{\omega_k}) + \tfrac{1}{2}\epsilon'^2 \bar{M}_{\omega\omega}(\xi_{\omega_k})$$

where the suffix ω on $\bar{M}$ denotes differentiation.

From (4), (5)

$$\bar{M}'_{k+1} \simeq \bar{M}(\xi_{k+1}) + \delta_1 \bar{M}_\omega(\xi_{\omega_k}) + \tfrac{1}{2}\delta_2 \bar{M}_{\omega\omega}(\xi_{\omega_k})$$

(6)

where

$$\delta_1 = (\lambda_i + \lambda_k)\epsilon' - \lambda_i \epsilon$$

(7)

and

$$\delta_2 = (\lambda_i + \lambda_k)\epsilon'^2 - \lambda_i \epsilon^2$$

(8)

Then the choice $\epsilon' = \lambda_i \epsilon/(\lambda_i + \lambda_k)$, i.e.

$$\omega' = \frac{\lambda_i \omega_i + \lambda_k \omega_k}{\lambda_i + \lambda_k}$$

(9)

yields $\delta_1 = 0$ and

$$\bar{M}'_{k+1} = \bar{M}(\xi_{k+1}) - \tfrac{1}{2}(\omega_i - \omega_k)^2 \frac{\lambda_i \lambda_k}{\lambda_i + \lambda_k} \bar{M}_{\omega\omega}(\xi_{\omega_k})$$

(10)

so that, to the first order in $(\omega_i - \omega_k)$, the information matrix is unchanged by the rounding-off procedure. Note that ω' is the weighted average of the two frequencies that it replaces.

From (5), a first order approximation is sufficient provided

$$|\epsilon| \ll \min_{i,j=1,\ldots,p} \left. |[\bar{M}(\xi_{\omega_k})]_{ij}| \middle/ |[\bar{M}_\omega(\xi_{\omega_k})]_{ij}| \right.$$

(11)

However, even if the replacement (9) is used whenever $|\omega_i - \omega_k| < \delta$ for some small *fixed* δ, this appears to improve the convergence rate and also reduce the final design index (see Appendix A).

4.11 CONCLUDING REMARKS

This chapter has considered in detail the problem of generating D-optimal designs of test signals for discrete-time systems. A class of algorithms has been proposed each yielding a sequence of design measures converging to a D-optimum. This completes the discussion of input design for discrete-time systems.

In the next chapter the analysis of Chapter 3 is extended to the case of continuous-time systems, leading to a discussion of the joint determination of optimal input signal and sampling rate.

Appendix A

COMPARISON OF SEQUENTIAL DESIGN ALGORITHMS

In this Appendix the rates of convergence of Algorithms 3 - 6
are compared for each of six process models. For each model the
starting design and stopping condition are the same for every
algorithm. The results of the sequential procedures are compared
with the D-optimal designs produced by a pattern search optimization
over a discrete design with a fixed number of frequencies.

The sequential round-off procedure of Section 10 is used if
$|\omega_i - \omega_j| < 0.1$, $i \neq j$. The cost function throughout is $\det \bar{\bar{M}}^{-1}$.

Example 1

See also Example (3.A.1).

$$\text{Model:} \qquad Y_k = (1 + 0.7z^{-1})u_k + \frac{1 + 0.3z^{-1}}{1 + 0.5z^{-1}} e_k \qquad (1)$$

Initial Design:	Frequency	Power Proportion				(2)
	1.0000	1.0000				

Final Designs:

(i) Pattern search method:

	Frequency	Power Proportion	N var $\hat{b}_0$	N var $\hat{b}_1$	Cost	(3)
	1.3804	1.0000	0.8673	0.8673	0.7253	

in the usual notation.

(ii) Sequential Procedures:

Every algorithm gives the same result as (i) using the stopping condition.

$$\left| \bar{M}(\xi_{k+1}) \right| \ / \ \left| \bar{M}(\xi_k) \right| \ < \ 1+\epsilon \tag{4}$$

where ϵ is taken to be 10^{-6}.

Also

Algorithm	Final Stage	
3	9	
4	3	(5)
5	12	
6	3	

Example 2

See also Example (3.A.2).

Model: $y_k = (1+0.7z^{-1})u_k + (1+0.3z^{-1})e_k$ (6)

The initial design is chosen as in (2).

Final Designs:

(i) Pattern Search Method:

Frequency	Power Proportion	N var $\hat{b}_0$	N var $\hat{b}_1$	Cost	(7)
2.1537	1.0000	1.0900	1.0900	0.8281	

(ii) Sequential Procedures:

Algorithm	Frequency	Power Proportion	Final Stage	(8)
3	1.0000	0.5428	2	
5	3.1102	0.4572	3	
4	1.0000	0.5416	3	
6	3.0788	0.4584	3	

Stopping Condition:

$$\varepsilon = 10^{-6}$$

The parameter variances and optimal cost are as in (7).

Example 3

See also Example (3.A.3)

Model: $$Y_k = \frac{1+0.3z^{-1}}{1+0.5z^{-1}} u_k + 0.1e_k \qquad (9)$$

Initial Design:

Frequency	Power Proportion	(10)
1.0000	0.5000	
2.0000	0.5000	

Final Designs:

(i) Pattern search method:

Frequency	Power Proportion	N var $\hat{a}_1$	N var $\hat{b}_0$	N var $\hat{b}_1$	Cost
1.3372	0.3976				
		0.0791	0.0131	0.1485	0.4449×10^{-5}
2.7984	0.6024				

(ii) Sequential Procedures:

Algorithm 3:	Frequency	Power Proportion		(12)
	1.0000	0.2105		
	2.0000	0.2105		
	2.6074	0.2394	Final Stage:	
	2.7586	0.1357	16	
	π	0.2039		

Algorithm 4:	Frequency	Power Proportion		(13)
	1.0000	0.2298		
	2.0000	0.2298	Final Stage:	
	2.8328	0.5403	16	

Algorithm 5:	Frequency	Power Proportion		(14)
	1.0000	0.1214		
	2.0000	0.3430		
	2.5761	0.2603	Final Stage:	
	π	0.2753	12	

Algorithm 6:	Frequency	Power Proportion		(15)
	0	0.0200		
	1.0000	0.1655		
	2.0000	0.2967	Final Stage:	
	2.8274	0.4675	10	
	π	0.0503		

Stopping condition:

$$\varepsilon = 10^{-6}$$

The parameter variances and optimal cost are as in (11).

Example 4

Model:
$$y_k = \frac{1-0.1z^{-1}}{1+0.5z^{-1}} u_k + \frac{0.1+0.07z^{-1}}{1+0.3z^{-1}} e_k \qquad (16)$$

Initial design as in Example 3.

Final Designs:

(i) Pattern search method:

Frequency	Power Proportion	N var $\hat{a}_1$	N var $\hat{b}_0$	N var $\hat{b}_1$	Cost
2.5007	2/3	0.0052	0.0150	0.0539	0.5637×10^{-7}
π	1/3				

(17)

(ii) Sequential procedures:

Algorithm 3:

(18)

Frequency	Power Proportion	
1.0000	0.0415	
2.0000	0.0415	
2.5077	0.4934	Final Stage:
2.7239	0.0882	94
2.8903	0.0266	
π	0.3088	

Algorithm 4:

Frequency	Power Proportion	(19)
1.0000	0.0255	
2.0000	0.0255	
2.5034	0.5861	Final Stage
2.9217	0.0733	139
π	0.2896	

Algorithm 5:

Frequency	Power Proportion	(20)
2.4339	0.5373	
2.5528	0.0296	
2.7332	0.1128	Final Stage:
π	0.3203	55

Algorithm 6:

Frequency	Power Proportion	(21)
2.3969	0.3974	
2.6043	0.2764	Final Stage:
π	0.3262	86

Stopping condition

$$\varepsilon = 10^{-4}$$

Algorithm	N var $\hat{a}_1$	N var $\hat{b}_0$	N var $\hat{b}_1$	Cost
3	0.0053	0.0153	0.0547	0.5832×10^{-7}
4	0.0053	0.0151	0.0543	0.5769×10^{-7}
5	0.0052	0.0148	0.0530	0.5693×10^{-7}
6	0.0052	0.0147	0.0527	0.5687×10^{-7}

$$(22)$$

Example 5

See also Example (3.B.1).

Model:
$$y_k = (1+0.3z^{-1})u_k + \frac{1}{1-0.06z^{-4}} e_k \qquad (23)$$

(i) Pattern search method:

Initial Design as in Example 3.

Final Design: (24)

Frequency	Power Proportion	N var $\hat{b}_0$	N var $\hat{b}_1$	Cost
0.7854 (=π/4)	½			
		0.8900	0.8900	0.7921
2.3562 (=3π/4)	½			

(ii) Sequential procedures:

Initial single frequency design as in Example 1.

Final Designs:

Algorithm 3: (25)

Frequency	Power Proportion	
0.7854	0.4415	
1.0000	0.0745	
2.3562	0.4314	Final Stage:
2.4819	0.0526	64

Algorithm 4: (26)

Frequency	Power Proportion	
0.7854	0.4159	
1.0000	0.0930	Final Stage:
2.3562	0.4911	62

Algorthm 5:   Frequency   Power Proportion   (27)

Frequency	Power Proportion	
0.7735	0.4970	Final Stage:
2.3562	0.5030	48

Algorithm 6:   Frequency   Power Proportion   (28)

Frequency	Power Proportion	
0.7854	0.5000	Final Stage:
2.3562	0.5000	14

Stopping Condition:

$$\varepsilon = 10^{-4}$$

Algorithm	N var $\hat{b}_0$	N var $\hat{b}_1$	Cost	(29)
3	0.8931	0.8931	0.7976	
4	0.8931	0.8931	0.7976	
5	0.8900	0.8900	0.7922	
6	0.8900	0.8900	0.7921	

Example 6

See also Example (3.B.2).

Model:
$$y_k = (1.0.5z^{-1}+0.06z^{-2})u_k + (1-0.23z^{-2}-0.08z^{-4}-0.07z^{-6})^{-1}e_k \tag{30}$$

(i) Pattern search method:

Initial Design: Frequency   Power Proportion   (31)

Frequency	Power Proportion
1	1/3
2	1/3
3	1/3

Final Design: (32)

Frequency	Power Proportion
0.6681	0.3270
1.5708	0.3459
2.4735	0.3270

$$N \text{ var } \hat{b}_0 = 0.8377 \qquad N \text{ var } \hat{b}_1 = 0.7800$$

$$N \text{ var } \hat{b}_2 = 0.8377 \qquad \text{Cost} = 0.5096$$

(ii) Sequential procedures:

Initial two-frequency design as in 3.

Final Designs:

Algorithm 3: (33)

Frequency	Power Proportion	
0.6588	0.3065	
1.0000	0.0367	
1.5708	0.3109	Final Stage:
2.0000	0.0367	171
2.4847	0.3091	

Algorithm 4: (34)

Frequency	Power Proportion	
0.6612	0.3067	
1.0000	0.0327	
1.5708	0.3187	Final Stage:
2.0000	0.0327	195
2.4828	0.3090	

Algorithm 5:	Frequency	Power Proportion		(35)
	0.6413	0.3211		
	1.5942	0.3707	Final Stage:	
	2.5199	0.3082	21	

Algorithm 6:	Frequency	Power Proportion		(36)
	0.6597	0.3266		
	1.5866	0.3557	Final Stage:	
	2.4961	0.3177	22	

Stopping condition:

$$\varepsilon = 10^{-4}$$

Algorithm	N var $\hat{b}_0$	N var $\hat{b}_1$	N var $\hat{b}_2$	Cost	(37)
3	0.8432	0.7867	0.8432	0.5219	
4	0.8445	0.7850	0.8445	0.5204	
5	0.8385	0.7823	0.8385	0.5131	
6	0.8371	0.7809	0.8371	0.5104	

Remarks

(1) In Examples 1 and 2 there exist D-optimal designs with a single frequency. In these cases the removal Algorithms 5 and 6 are not useful as the addition algorithms can choose $\alpha_k = 1$ thus eliminating the previous design. Note that for $p > 2$ the possibility of removal of frequencies gives substantial improvement.

(2) Examples 2 and 3 are models satisfying the hyperplane condition (3.5.3). In these cases the D-optimal $\bar{M}(\xi^*)$ lies in the relative interior of M. Then any starting design can be used to initiate a simplex enclosing the D-optimum. In this sense there is no 'bad' initial design ξ_0 and this leads to relatively few stages before the algorithms converge.

(3) In contrast, Example 4 corresponds to condition (iv) of Theorem (3.2), i.e. only principal D-optimal designs are possible and the hyperplane condition is not satisfied. This imposes severe restrictions on the final design and is reflected in the large number of steps taken by the algorithms.

(4) Examples 5 and 6 indicate the advantage of using sequential procedures. The initial designs have $[(p+1)/2]$ frequencies but the final D-optimal designs require p frequencies (after removing the remnants of ξ_0).

(5) In general the sequential algorithms are slow to converge and, for $p > 2$, there appears to be little advantage in using Algorithms 4 or 6. Frequency removal substantially improves the convergence rate and therefore Algorithm 5 appears to be the best out of the sequential procedures considered when $p > 2$.

Chapter 5

CONTINUOUS-TIME SYSTEMS

5.1   INTRODUCTION

In this chapter, the Tchebycheff system approach to input design developed in Chapter 3 for the discrete-time case is extended to continuous-time systems with p estimable parameters.

Initially, the problem of infinite frequency bands is avoided by assuming a finite cut-off frequency ω_c. It is shown (Section 5) that the set of information matrices corresponding to normalised input power can be represented as a set $M^{(p)}(\omega_c)$, a subset of a closed convex cone $M_c^{(p)}(\omega_c)$, induced in R^p by a complete Tchebycheff system on the interval $[0, \omega_c]$. This leads to a number of sufficient conditions for the existence of optimal canonical representations (Section 7). The case of normalised output power is briefly considered in Section 9.

The arbitrariness of ω_c leads to the possibility of a Φ-optimum occurring on the cone boundary with design index $I(\omega_c; \xi) = (p-1)/2$. It is shown (Section 6) that this does not occur for $q-r < n-m$ provided ω_c is sufficiently large. In the case $n-m = q-r$ (Section 8), ω_c can be taken to its infinite limit and the finite-interval theory carries over unchanged.

In the appendix, the problem of using 'optimal' inputs based on estimated (approximate) parameter values is discussed. It is shown through a simple example that such inputs should be used with care.

5.2   MODEL STRUCTURE

The model considered here is analogous to that constructed in Section (2.2) for a discrete-time system. The model is of the form

$$y(t) = \frac{B(s)}{A(s)} u(t-\tau) + \frac{D(s)}{C(s)} e(t) \tag{1}$$

$$t \in [t_0, t_f]$$

in the usual notation, where τ is a known time delay and $e(t)$ is a unit variance Gaussian white noise process. The parameterisation is given by

$$A(s) = 1 + a_1 s + \ldots + a_n s^n$$

$$B(s) = b_0 + b_1 s + \ldots + b_m s^m \tag{2}$$

$$C(s) = 1 + c_1 s + \ldots + c_q s^q$$

$$D(s) = d_0 + d_1 s + \ldots + d_r s^r$$

$$\beta = (\Theta^T, c_1, \ldots, c_q, d_0, \ldots, d_r)^T$$

$$\Theta = (a_1, \ldots, a_n, b_0, \ldots, b_m)^T$$

$$\in R^p \text{ where } p = m+n+1$$

The polynomials A, B, C, D have no zeros in the closed right-half plane.

It is also assumed that the rational part of the system transfer function $u \to y$ is regular, i.e.

$$m \leq n \tag{3}$$

and that the noise transfer function $e \to y$ satisfies

$$r \leq q \tag{4}$$

The inclusion of the possibility of equality in (4) can be justified for estimation purposes [S5, p. 207ff].

The experiment design problem considered initially is that of choosing the input signal in order to achieve the greatest accuracy in estimating the process parameter vector β from input-output data.

5.3 FREQUENCY DOMAIN APPROACH

Following the development in the discrete-time case (Chapter 2) it is again assumed that the experiment time is long (i.e. $T \triangleq t_f - t_0$ is large compared with the largest time constant) and that an input or output power constraint is imposed. Thus the frequency domain approach is suitable.

The asymptotic per unit time information matrix $\bar{M}$ is given by [V5, p. 167ff]

$$[\bar{M}_\beta]_{ik} = \frac{1}{2\pi} \int_{-\infty}^{\infty} \{\frac{\partial\mu(-j\omega)}{\partial\beta_i} \frac{\partial\mu(-j\omega)}{\partial\beta_k} \frac{1}{\psi(\omega)} + \frac{1}{2} \frac{\partial\psi(\omega)}{\partial\beta_i} \frac{\partial\psi(\omega)}{\partial\beta_k} \frac{1}{[\psi(\omega)]^2} \} \, d\omega$$

$$i, k = 1, 2, \ldots, p+r+q+1 \tag{1}$$

where $\mu(j\omega)$ is the complex Fourier transform of the mean of the process output, i.e.

$$\mu(s) = e^{-s\tau} \frac{B(s)}{A(s)} u(s) \tag{2}$$

and $\psi(\omega)$ is the spectral density of the output, i.e.

$$\psi(\omega) = \frac{D(j\omega)D(-j\omega)}{C(j\omega)C(-j\omega)} \tag{3}$$

It follows immediately from (1)-(3) that $\bar{M}_\beta$ is of the form

$$\bar{M}_\beta = \begin{bmatrix} \bar{M} & 0 \\ 0 & \bar{R} \end{bmatrix} \tag{4}$$

where $\bar{R}$ corresponds to the noise parameters and is independent of the input. The pxp submatrix $\bar{M}$ corresponding to the system parameters can be written in the form

$$\bar{M} = \text{Re} \int_0^\infty h(j\omega) h^*(j\omega) d\xi(\omega) \tag{5}$$

[c.f. Section (2.6)] where the column p-vector h is given by

$$h_i(s) = -e^{s\tau} \frac{CB}{DA^2}(s) s^i \qquad i = 1, \ldots, n \tag{6}$$

$$= e^{s\tau} \frac{C}{DA}(s) s^{(i-n-1)} \qquad i = n+1, \ldots, p$$

and $\xi(\omega)$ is the (one-sided) cumulative power distribution function of the input signal.

The normalised input power constraint corresponds to the condition

$$\int_0^\infty d\xi(\omega) = 1 \tag{7}$$

The similarity between the expressions obtained above and those developed for the discrete-time case is to be expected. However, an important difference is the replacement of the closed interval $\{\omega \mid 0 \leq \omega \leq \pi\}$ by the semi-infinite interval $\{\omega \mid 0 \leq \omega < \infty\}$. The problem of closing the interval (by allowing improper input design measures with positive weight at $\omega = \infty$) is dealt with later (Section 8).

The input spectrum can always be confined approximately to some finite interval $[0, \omega_c]$ where ω_c is a suitable cut-off frequency and this is assumed below.

5.4 PROPERTIES OF $\bar{M}$

Assuming that

$$d\xi(\omega) = 0 \quad \text{for all } \omega > \omega_c \tag{1}$$

the expression for the information matrix $\bar{M}$ is

$$\bar{M} = \text{Re} \int_0^{\omega_c} h(j\omega) h^*(j\omega) d\xi(\omega) \tag{2}$$

where h is given by equation (3.6).

Definition 1

Denote by $\Xi_1(\omega_c)$ the set of ξ for which

$$\int_0^{\omega_c} d\xi(\omega) = 1 \tag{3}$$

and denote by $\mathcal{D}_1(\omega_c)$ the subset of $\Xi_1(\omega_c)$ for which ξ has a purely discrete spectrum.

Definition 2

If ξ has a purely discrete spectrum define a *design index* $I(\omega_c; \xi)$ as the number of design frequencies in $[0, \omega_c]$ where the end frequencies 0 and ω_c are each counted as one half.

In addition, define a *modified design index* (MDI) $\hat{I}(\omega_c; \xi)$ as above except that only the end frequency $\omega = 0$ is given a half-count.

The need to introduce $\hat{I}(\omega_c; \xi)$ is dictated by the arbitrariness of ω_c. Note that

$$\text{rank } [h(j\omega)h^*(j\omega)] = 1 \qquad \text{iff } \omega = 0$$

$$(4)$$

$$= 2 \qquad \text{for all } \omega \in (0, \omega_c]$$

and this leads to:

Theorem 1

The information matrix $\bar{M}$ has the following properties:

(i) $\bar{M}$ is a real, symmetric non-negative definite matrix;

*(ii) $\bar{M}$ is nonsingular iff the design measure ξ satisfies the condition

$$\hat{I}(\omega_c;\xi) \geq p/2.$$

Proof

(i) The result follows directly from (2);

(ii) The proof is exactly as for Theorem (2.8.2) except that (4) implies

that the design index is to be replaced throughout the proof by the MDI

$\hat{I}(\omega_c;\xi)$. #

Remark 1

Note that $\bar{M}$ is nonsingular if $\hat{I}(\omega_c;\xi) = p/2$, corresponding to

$I(\omega_c;\xi) = (p-1)/2$ (< p/2) if ω_c belongs to the spectrum of ξ.

Definition 3

Let $M(\omega_c)$ denote the set of information matrices corresponding to

the set $\Xi_1(\omega_c)$ of design measures.

It then follows that [c f. Theorems (3.2.1) and (3.3.1)]:

Theorem 2 [K8, M3, P3]

(i) The set $M(\omega_c)$ is compact and is the convex hull of the subset of

$M(\omega_c)$ corresponding to single frequency designs;

(ii) If $\xi_1 \in \Xi_1(\omega_c)$, then there exists $\xi_2 \in D_1(\omega_c)$ whose spectrum contains

no more than p+1 points and such that $\bar{M}(\xi_1) = \bar{M}(\xi_2)$. If $\bar{M}(\xi_1)$ lies on the

boundary of $M(\omega_c)$ then no more than p points are required. #

The information matrix $\bar{M}$ can again be expressed in the form

$$\bar{M} = \sum_{i=1}^{p} \alpha_i L_i \tag{5}$$

where $L_1, \ldots, L_p$ are constant matrices,

$$\alpha_i = \int_0^{\omega_c} v_i(\omega) \, d\xi(\omega) \qquad i = 1, \ldots, p \tag{6}$$

and

$$v_i(\omega) = f(\omega) \omega^{2(i-1)} \qquad i = 1, \ldots, p \tag{7}$$

where

$$f(\omega) = \frac{C}{DA^2}(j\omega) \frac{C}{DA^2}(-j\omega) \tag{8}$$

The input design problem can again be approached using Tchebycheff system theory.

5.5 COMPLETE TCHEBYCHEFF SYSTEMS

Definition 1

$\{v_i\}_1^p$ is a complete Tchebycheff system (CT-system) on $[a,b]$ if $\{v_i\}_1^k$ is a T-system on $[a,b]$ for $k = 1, 2, \ldots, p$.

* ### Theorem 1

$\{v_i\}_1^p$ is a CT-system on $[0,\delta]$ for all $\delta > 0$.

Proof

From (4.7)

$$V\left(\begin{array}{c} 1, \ 2, \ \ldots, \ k \\ \omega_1, \ \omega_2, \ \ldots, \ \omega_k \end{array} \right) = \left| \begin{array}{ccc} f(\omega_1) & \cdots & f(\omega_k) \\ f(\omega_1)\omega_1^2 & \cdots & f(\omega_k)\omega_k^2 \\ \vdots & & \vdots \\ f(\omega_1)\omega_1^{2(k-1)} & \cdots & f(\omega_k)\omega_k^{2(k-1)} \end{array} \right|$$

$$= [\prod_{i=1}^{k} f(\omega_i)]\prod_{1\leq s<t\leq k}(\omega_t^2 - \omega_s^2) \tag{1}$$

where $0 \le \omega_1 < \omega_2 < \ldots < \omega_k \le \omega_c$ and $f(\omega)$ is continuous and positive

for all finite ω.

It follows that

$$V \begin{pmatrix} 1, & 2, & \ldots, & k \\ \omega_1, & \omega_2, & \ldots, & \omega_k \end{pmatrix} > 0 \quad k = 1, 2, \ldots, p$$

q.e.d.   #

The CT-system $\{v_i\}_1^p$ induces a closed convex cone $M_c^{(p)}(\omega_c)$ in R^p,

i.e.

$$M_c^{(p)}(\omega_c) = \{ \underline{x} \epsilon R^p \mid \underline{x} = \int_0^{\omega_c} \underline{v}(\omega) \, d\xi(\omega), \; \xi \epsilon \Xi(\omega_c) \} \tag{2}$$

where $\Xi(\omega_c)$ denotes the set of all nondecreasing right continuous

functions of bounded variation on $[0, \omega_c]$. The set $M^{(p)}(\omega_c)$ defined as

in (2) with $\Xi_1(\omega_c)$ replacing $\Xi(\omega_c)$ is isomorphic to $M(\omega_c)$ and is a subset

of the cone.

5.6   THE CONE BOUNDARY

Design measures exist which give rise to points on the boundary

of the cone, Bd $M_c^{(p)}(\omega_c)$, and yet correspond to persistently exciting

inputs. This differs from the discrete-time case [c.f. Result (3.6.1)].

The spectra of such designs satisfy one of the following conditions:

Case (i)

   p even

$$0 < \omega_1 < \omega_2 < \ldots < \omega_{p/2} = \omega_c \tag{1a}$$

Case (ii)

   p odd

$$0 = \omega_1 < \omega_2 < \ldots < \omega_{(p+1)/2} = \omega_c \tag{1b}$$

In both cases the MDI is p/2 and therefore, from Theorem (4.1), gives rise to a nonsingular information matrix. However, the design index is $(p-1)/2$ and therefore gives rise to a point on the boundary of the cone [see Main Appendix].

Result 1

Let $\underline{x} \in \text{Bd } M_c^{(p)}(\omega_c)$. Then $\underline{x}$ has a unique representation.

Proof

See Main Appendix.   #

The question arises: Under what conditions can a Φ-optimal point $\underline{x}$ lie in $[\text{Bd } M_c^{(p)}(\omega_c)] \cap [\text{Bd } M^{(p)}(\omega_c)]$?

If such a Φ-optimum exists, it corresponds to a unique design measure ξ^* (Result 1) whose spectrum is given by (1a) or (1b). Fears that ξ^* may not satisfy the input constraint are allayed by:

* Result 2

Let $\underline{x} \in [\text{Bd } M_c^{(p)}(\omega_c)] \cap [\text{Bd } M^{(p)}(\omega_c)]$. Then $\underline{x}$ corresponds to a unique design measure which lies in $\Xi_1(\omega_c)$.

Proof

Result 1 ensures that $\underline{x}$ corresponds to a unique design measure ξ. Hence if $\xi \notin \Xi_1(\omega_c)$, then $\underline{x} \notin M^{(p)}(\omega_c)$ contrary to assumption. The result follows.   q.e.d.

Consider the following example:

Example 1

$A(s) = 1 + s\tau$; $B(s) = k$, i.e. $p = 2$. Then the average information matrix $\bar{M}(\omega)$, corresponding to a single frequency design measure in $\Xi_1(\omega_c)$, is given by

$$\bar{M}(\omega) = [\psi(\omega)(1+\omega^2\tau^2)^2]^{-1} \begin{bmatrix} k^2\omega^2 & -k\omega^2\tau \\ \\ -k\omega^2\tau & 1+\omega^2\tau^2 \end{bmatrix} \tag{2}$$

where $\psi(\omega) = |D(j\omega)/C(j\omega)|^2$ is the noise spectral density. Note that $\bar{M}(0)$ is singular but that $\bar{M}(\omega_c)$ is not. Then

$$\bar{M}^{-1}(\omega) = \psi(\omega)\,(\frac{1+\omega^2\tau^2}{k\omega})^2 \begin{bmatrix} 1+\omega^2\tau^2 & k\omega^2\tau \\ k\omega^2\tau & k^2\omega^2 \end{bmatrix} \tag{3}$$

and

$$\det \bar{M}^{-1}(\omega) = [\psi(\omega)(1+\omega^2\tau^2)^2/(k\omega)]^2 \tag{4a}$$

$$\text{trace } [W\bar{M}^{-1}(\omega)] = \psi(\omega)(1+\omega^2\tau^2)^2(\alpha+\beta\omega^2)/(k\omega)^2 \tag{4b}$$

where W is a positive definite matrix and α, β are positive numbers dependent on its elements.

It follows that

$$\lim_{\omega\to\infty} \det \bar{M}^{-1}(\omega) = 0 \qquad \text{iff } q-r > 1 \tag{5a}$$

$$\lim_{\omega\to\infty} \text{trace } [W\bar{M}^{-1}(\omega)] = 0 \qquad \text{iff } q-r > 2 \tag{5b}$$

where q, r are the orders of the polynomials C, D respectively. Therefore if $q-r > 1$, $\omega = \omega_c$ is the best choice of design frequency (in the sense of D-optimality) provided that ω_c is large enough.   #

In Example 1, the condition $q-r > 1$ can be interpreted as implying a faster roll-off of the noise transfer function compared with the system transfer function, i.e. the number of zeros at infinity of D(s)/C(s) exceeds that for B(s)/A(s). This is also the condition for $v_i(\omega)$ to be unbounded on the positive real line for some i.

In general the condition for $v_i(\omega)$ to be unbounded on the positive real line for some i is that $q-r > n-m$ but Example 1 shows that this may not be sufficient to ensure that

$$\lim_{\omega_\ell \to \infty} \Phi[\bar{M}(\xi)] \to 0 \tag{6}$$

for all Φ of interest, where ξ has a discrete spectrum whose highest

frequency is ω_ℓ.

However, if $q-r < n-m$, then $\lim_{\omega \to \infty} v_i(\omega) = 0$, $i = 1, 2, \ldots, p$ and

therefore, for sufficiently high values of the cut-off, the inclusion

of ω_c in a design spectrum yields only small contributions to the

elements of $\bar{M}(\xi)$. In particular, if the design spectrum is given by

(1a) or (1b) then the information matrix will be almost singular and

far from Φ-optimal, i.e.

* Result 3

If $q-r < n-m$ and ω_c is sufficiently large then the Φ-optimum lies

in Int $M_c^{(p)}(\omega_c)$.

The case $n-m = q-r$ is special and will be dealt with later.

5.7 THE CONE INTERIOR

All interior points of the cone correspond to nonsingular

information matrices. The following results are of importance and are

proved as in Chapter 3, with the added restriction that the model orders

satisfy

$$m \leq n, \qquad r \leq q \tag{1}$$

* Result 1

The moment space $M^{(p)}(\omega_c)$ is a hyperplane in R^p iff

$$m = n, \qquad q = 0 = r \tag{2}$$

Proof

See Theorem (3.5.2) and use (1).   #

Defining canonical and principal representations as in Section (3.6)

(ω_c replacing the end frequency π) it follows that:

* **Result 2**

If $M^{(p)}(\omega_c)$ is a hyperplane and $\underline{x}^0 \in M^{(p)}(\omega_c) \cap \text{Int } M_c^{(p)}(\omega_c)$ then, for any ω^* in $(0,\omega_c)$, there exists a unique canonical representation ξ^0 of $\underline{x}^0$ involving ω^* and such that $\xi^0 \in \Xi_1(\omega_c)$.

Proof

See Theorem (3.6.4).   #

* **Result 3**

If $M^{(p)}(\omega_c)$ is not a hyperplane in R^p and $\underline{x}^0 \in [\text{Bd } M^{(p)}(\omega_c)] \cap [\text{Int } M_c^{(p)}(\omega_c)]$, then every design measure ξ in $\Xi_1(\omega_c)$ is discrete with index satisfying the inequality

$$I(\omega_c;\xi) \leq \tfrac{1}{2} \max (p+q-1,2n+r) \tag{3}$$

Proof

See Theorem (3.7.1).

Remark 1

Another form of (3) is

$$I(\omega_c;\xi) \leq n + \tfrac{1}{2}q - \tfrac{1}{2} \min (n-m,q-r) \tag{4}$$

If $q-r < n-m$, then Result (6.3) ensures that the Φ-optimum lies in the cone interior and the upper bound in (4) is $n + r/2$ for such a point. #

* **Result 4**

If $\underline{x}^0 \in [\text{Bd } M^{(p)}(\omega_c)] \cap [\text{Int } M_c^{(p)}(\omega_c)]$ and $M^{(p)}(\omega_c)$ is not a hyperplane in R^p, then a sufficient condition that every representation in $\Xi_1(\omega_c)$ of $\underline{x}^0$ is canonical is that one of the following conditions is satisfied:

(i) $m = n-1;$ $r = 0 = q$

(ii) $m = n-2;$ $r = 0 = q$

(iii) $m = n;$ $r = 0$ or $1;$ $q = 1$

(iv) $m = n-1;$ $r = 0;$ $q = 1$

(v) $m = n-1;$ $r = 1 = q$ (5)

(vi) $m = n-2;$ $r = 0;$ $q = 1$

(vii) $m = n;$ $r \leq 2;$ $q = 2$

(viii) $m = n-1;$ $r = 0$ or $1;$ $q = 2$

(ix) $m = n-2;$ $r = 0;$ $q = 2$

In cases (i), (iii) and (iv) only principal representations of $\underline{x}^0$ are possible.

Proof

See Theorem (3.7.2) and use (1).   #

Example 1

Zarrop et al [Z3] consider D-optimal design measures for the model

$$y(s) = \frac{k}{1+s\tau} u(s) + \frac{1}{1+as} e(s) \qquad (6)$$

using Payne's result [Theorem (4.2)(ii)]. The average information matrix $\bar{M}$ is given by

$$\bar{M} = \int_0^{\omega_c} \begin{bmatrix} 1+\tau^2\omega^2 & -k\tau\omega^2 \\ -k\tau\omega^2 & k^2\omega^2 \end{bmatrix} (1+a^2\omega^2)/(1+\tau^2\omega^2)^2 \, d\xi(\omega) \qquad (7)$$

but at most two design frequencies ω_1, ω_2 are required with respective weights λ, $1-\lambda$ to satisfy the constraint of normalised input power.

Hence an optimisation problem in R^3 has to be solved. Some tedious manipulation of the stationarity conditions leads to $\omega_1 = \omega_2$ so that only one frequency is required and its D-optimal value ω_0 is found to satisfy

$$a^2 \tau^2 \omega_0^4 + 3(\tau^2 - a^2)\omega_0^2 - 1 = 0 \qquad (8)$$

In the white noise case (a = 0), the solution of (8) is $\omega_0 = (\tau\sqrt{3})^{-1}$.

Alternatively: the orders of the polynomials in (6) satisfy

(i) n-m = q-r, (ii) m = n-1, r = 0, q = 1. Condition (i) indicates the possibility that the D-optimum may lie on the boundary of the cone $M_c^{(2)}(\omega_c)$ with design index (p-1)/2 = ½ (Section 6). If the D-optimum lies in the cone interior, then condition (ii) implies that the optimal design index is p/2 = 1 (Result 4).

Therefore, in either case, only one frequency is necessary.

If a = 0, then n-m > q-r and the optimum lies in the cone interior [Result (6.3)] with design index 1 [Condition (5)(i)].

5.8   THE CASE n-m = q-r

Returning to the discussion of Section 3, consider the case in which the cut-off frequency ω_c becomes infinite. The convex cone $M_c^{(p)}$ defined by

$$M_c^{(p)} = \{\underline{x} \in R^p \,|\, \underline{x} = \int_0^\infty \underline{v}(\omega)\, d\xi(\omega), \; \xi \in \Xi\} \qquad (1)$$

where Ξ denotes the set of all nondecreasing right continuous functions of bounded variation on $[0,\infty)$ is not in general closed and this may lead to the Φ-optimum lying outside $M_c^{(p)}$.

Closure of the cone can be achieved by the introduction of improper design measures allowing $\omega = \infty$ to be given positive weight. The simplest case is the following [K8, p. 147]:

Definition 1

A T-system $\{v_i\}_1^p$ is said to be of Type I if the following conditions are satisfied:

(i) $\lim\limits_{\omega \to \infty} v_i(\omega)$ is finite, $i = 1, 2, \ldots, p$

(2)

(ii) $\lim\limits_{\omega_p \to \infty} V\begin{pmatrix} 1, & 2, & \ldots, & p \\ \omega_1, & \omega_2, & \ldots, & \omega_p \end{pmatrix} > 0 \quad 0 \le \omega_1 < \ldots < \omega_p < \infty$

in the usual notation.

Condition (2)(i) implies that $n-m \ge q-r$. Condition (2)(ii) ensures that the Tchebycheff property is preserved in the limit and leads to $n-m \le q-r$. It follows that:

* ## Result 1

The CT- system $\{v_i\}_1^p$ is a Type I system iff $n-m = q-r$.

In this case

$$\lim\limits_{\omega \to \infty} v_i(\omega) = 0 \qquad i = 1, 2, \ldots, p-1$$

(3)

$$\lim\limits_{\omega \to \infty} v_p(\omega) = v \triangleq (\frac{c_q}{d_r a_n^2})^2 > 0$$

using the usual notation for the polynomial coefficients.

Thus

$$\{v_1(\infty), v_2(\infty), \ldots, v_p(\infty)\} = \{0, 0, \ldots, 0, v\}$$

and the positive ray through this point of R^p can be added to the cone $M_c^{(p)}$ to give its closure.

In the notation of the previous section, $M_c^{(p)}(\infty)$ can be given a meaning by identifying it with the closure of $M_c^{(p)}$.

By carrying out the above compactification, the analysis of the cone $M_c^{(p)}(\infty)$ reduces at once to that for the finite interval case.

Further:

* Result 2

If n-m = q-r and the information matrix $\bar{M}$ is written in the form

$$\bar{M} = \int_0^\infty M(\omega)\,d\xi\,(\omega) \qquad (4)$$

then

$$\text{rank } M(\infty) = 1 \qquad (5)$$

Proof

Using (3.5) and (3.6) it is easily shown that, if n-m - q-r, then the only non-zero elements of $M(\infty)$ are

$$M(\infty)_{nn} = \nu b_m^2(-1)^{p-1}$$

$$M(\infty)_{np} = M(\infty)_{pn} = -\nu b_m a_n (-1)^{p-1}$$

$$M(\omega)_{pp} = \nu a_n^2(-1)^{p-1}$$

so that $M(\infty)$ can be written as the outer product

$$M(\infty) = (-1)^{p-1}\nu \alpha^T \alpha \qquad (6)$$

where $\alpha_n = -b_m$ and $\alpha_p = a_n$ are the only non-zero elements of the p-vector α. The result follows immediately.   #

Thus, in this case, the rank problems introduced by the arbitrariness of a finite cut-off disappear and all the results developed for the discrete-time case (Chapters 2-4) carry over completely (after the necessary notational changes).

Note that the conditions $r \leq q$, $m \leq n$, imposed in the continuous-time case, can lead to more specific statements as in Section 7. In particular, using Result (7.1), it follows that:

* Result 3

If $M^{(p)}(\omega_c)$ is a hyperplane in R^p, then the CT-system $\{v_i\}_1^p$ is of Type I.

Proof

If $M^{(p)}(\omega_c)$ is a hyperplane, then $m = n$, $q = 0 = r$ and thus $m-n = q-r$.   q.e.d.   #

The following examples illustrate a number of the preceding results.

Example 1

Consider the system $(p = 3)$

$$y(s) = \frac{1+s}{1+2s} \, u(s) + \frac{1+0.2s}{1+5s} \, e(s) \tag{7}$$

Then $m-n = q-r$ and the CT-system $\{v_i\}_1^3$ is of Type I where

$$v_i(\omega) = \frac{(1+25\omega^2)\,\omega^{2(i-1)}}{(1+0.04\omega^2)\,(1+4\omega^2)^2} \qquad i = 1, 2, 3 \tag{8}$$

Using Result (7.4), the D-optimal design has index 3/2, i.e. two frequencies including either 0 or ∞ with weight 1/3 [Theorem (4.2.1)]. Thus the design problem reduces to two optimisations in R^1 (or one in R^2) yielding the D-optimum:

frequency	power proportion
0.777	2/3
∞	1/3

(9)

5.8

149

Also

$$\det \bar{M}^{-1} = 0.158$$

and

$$\text{var } \hat{a}_1 = 26.08T^{-1}$$

$$\text{var } \hat{b}_0 = 1.14T^{-1} \tag{10}$$

$$\text{var } \hat{b}_1 = 6.67T^{-1}$$

where T is the experiment time.

Example 2

$$p = 3$$

$$y(s) = \frac{1+s}{1+2s} u(s) + \frac{1}{1+5s} e(s) \tag{11}$$

In this case the D-optimum may lie on the cone boundary or else Result (7.4) again applies. In either case no more than two frequencies are necessary and the D-optimum is

frequency	power proportion
0.494	0.383
10^4	0.617

$$\det \bar{M}^{-1} = 0.131 \times 10^{-7}$$

$$\text{var } \hat{a}_1 = 26.51T^{-1}$$

$$\text{var } \hat{b}_0 = 1.44T^{-1} \tag{12}$$

$$\text{var } \hat{b}_1 = 6.63T^{-1}$$

Here $\omega_c = 10^4$ and n-m < q-r so that, although the D-optimum is a principal design, the rank condition leading to Theorem (4.2.1) is violated and the power proportion for ω_c is not double that for the frequencies in $(0,\omega_c)$ as in Example 1.

Example 3

   p = 3

$$y(s) = \frac{1+s}{1+2s} u(s) + e(s) \qquad (13)$$

In this case the CT-system is of Type I and, in addition, $M^{(3)}(\infty)$ is a hyperplane in R^3. Hence, any chosen frequency ω^* can be used as part of a two-frequency ϕ-optimal design [Result (7.2)]. If $\omega^* = 0$ or ∞ as in I or II respectively then the design is principal.

The principal D-optimum designs are

frequency	power proportion
0	1/3
0.866	2/3

I

frequency	power proportion
0.289	2/3
∞	1/3

II

yielding det $\bar{M}^{-1}$ = 4096 and

   var $\hat{a}_1$ = 256T^{-1}

   var $\hat{b}_0$ = 3T^{-1} $\qquad\qquad$ (14)

   var $\hat{b}_1$ = 108T^{-1}

5.9   OUTPUT POWER CONSTRAINT

The above analysis carries over in a straightforward manner to the case of constrained output power. Following the discussion in Sections (2.7) and (3.8), the information matrix $\bar{M}$ can be written in the form

$$\bar{M} = \text{Re} \int_0^{\omega_c} \tilde{h}(j\omega)\tilde{h}*(j\omega)\,d\eta(\omega) \tag{1}$$

where $\eta \in \Xi_1(\omega_c)$,

$$\tilde{h}(s) = A(s)h(s)/B(s) \tag{2}$$

and h(s) is given by Equations (3.6).

The following results are of interest:

* Result 1

The system $\{\tilde{v}_i\}_1^p$ is a CT-system on $[0,\infty)$, i.e. on $[0,\omega_c]$ for all $\omega_c > 0$, where

$$\tilde{v}_i(\omega) = \tilde{f}(\omega)\omega^{2(i-1)} \qquad i = 1, \ldots, p \tag{3}$$

$$\tilde{f}(\omega) = \frac{C}{DAB}(j\omega)\frac{C}{DAB}(-j\omega) \tag{4}$$

Notation

Notation used is that of Section (3.8) apart from ω_c argument.

* Result 2

The following statements are equivalent:

(i) $\tilde{v}_i(\omega)$ is bounded for real ω, $i = 1, \ldots, p$

(ii) $\{\tilde{v}_i\}_1^p$ is of Type I

(iii) $r = q$

Remark 1

The function $\tilde{v}_i$ is unbounded on $[0,\infty)$ for some i if $r < q$. Hence, unless the process output has a white noise component, the cut-off ω_c may appear in any Φ-optimal design.

* Result 3

The moment space $\tilde{M}^{(p)}(\omega_c)$ is a hyperplane in R^p iff $r = q = 0$.

It follows from Result (7.1) that if $M^{(p)}(\omega_c)$ is a hyperplane then so is $\tilde{M}^{(p)}(\omega_c)$.

Remark 2

Result (7.2) carries over unchanged, except that $\tilde{M}$ replaces M.

* Result 4

If $\tilde{M}^{(p)}(\omega_c)$ is not a hyperplane in R^p and $\underline{x}^0 \in [\text{Bd } M^{(p)}(\omega_c)] \cap [\text{Int } M_c^{(p)}(\omega_c)]$, then every design measure η in $\Xi_1(\omega_c)$ representing $\underline{x}^0$ is discrete with index satisfying the inequality

$$I(\omega_c;\eta) \leq \tfrac{1}{2}(p+q-1) \tag{5}$$

Proof

Use Theorem (3.8.3) and the inequality $r \leq q$.   #

* Result 5

Under the conditions of Result 4, every representation in $\Xi_1(\omega_c)$ of $\underline{x}^0$ is canonical if

$$q \leq 2 \tag{6}$$

Only principal representations are possible if $q = 1$.

5.10 CONCLUDING REMARKS

The results of this chapter demonstrate that the Tchebycheff system framework erected in Chapter 3 to analyse the input design problem for

discrete-time systems can be suitably extended and modified to cope

with continuous-time systems.

Appendix A

COST SENSITIVITY UNDER PARAMETER UNCERTAINTY

The determination of optimal input signals requires knowledge of a priori parameter values. The true parameters, however, are not known - if they were, there would be no design problem to solve. It is desirable to determine how satisfactory the 'optimal' inputs are when the design parameters are different from the true parameters.

Gupta et al [G5] have carried out a sensitivity analysis for a specific flight testing problem and find that the optimal inputs are not very sensitive to estimation errors of up to 50%. In general, however, optimal designs should be used judiciously as is illustrated by the following example.

Consider the continuous-time system (p = 2)

$$y(s) = \frac{b}{1+as} u(s) + \frac{1}{1+cs} e(s)$$

in the usual notation, where a, b, c denote true parameter values. The elements of $\bar{M}$, the average information matrix per unit time, are given by

$$\bar{M}_{aa} = \pi^{-1} \int_0^\infty \frac{b^2 \omega^2 (1+c^2 \omega^2)}{(1+a^2 \omega^2)^2} d\xi(\omega) \qquad (1)$$

$$\bar{M}_{ab} = -a\bar{M}_{aa}/b \qquad (2)$$

$$\bar{M}_{bb} = \pi^{-1} \int_0^\infty \frac{1+c^2 \omega^2}{1+a^2 \omega^2} d\xi(\omega) \qquad (3)$$

The determinant criterion yields

$$det \ \bar{M}^{-1} = [\bar{M}_{aa}\bar{M}_{bb} - \bar{M}_{ab}^2]^{-1} = b^2 [\bar{M}_{aa}(b^2\bar{M}_{bb} - a^2\bar{M}_{aa})]^{-1} \qquad (4)$$

The cost minimum can be achieved with a single frequency input. For

input frequency ω

$$\det \bar{M}(\omega)^{-1} = \frac{(1+a^2\omega^2)^4}{b^2\omega^2(1+c^2\omega^2)^2} \tag{5}$$

The frequency value ω_0 that minimizes the cost when a, b, c are replaced

by estimated values $\hat{a}$, $\hat{b}$, $\hat{c}$ satisfies the quartic

$$(\hat{c}\hat{a})^2\omega^4 + 3(\hat{a}^2-\hat{c}^2)\omega^2 - 1 = 0 \tag{6}$$

i.e.

$$\omega_0 = [3(\hat{c}^2-\hat{a}^2)+\sqrt{\{9(\hat{c}^2-\hat{a}^2)^2+(2\hat{c}\hat{a})^2\}}]^{\frac{1}{2}}/(\hat{c}\hat{a}\sqrt{2}) \tag{7a}$$

$$\text{if } \hat{c} \neq 0$$

$$= 1/(\hat{a}\sqrt{3}) \qquad\qquad \text{if } \hat{c} = 0 \tag{7b}$$

If $\hat{a}$ = a and $\hat{c}$ = c, then ω_0 = ω^*, the optimal frequency. The frequency

ω_0 is independent of $\hat{b}$ and therefore only cost sensitivity with respect

to $\hat{a}$, $\hat{c}$ is relevant.

Tables 1-4 depict percentage increases of the cost value over the

minimum when $\hat{a}$, $\hat{c}$ are allowed to deviate independently from their true

values. It is assumed that b = 1 throughout and the following four

cases are considered:

(i) a = 1, c = 10   (ii) a = 1, c = 1

(iii) a = 1, c = 0.9   (iv) a = 1, c = 0

(In case (iv) for which c = 0, $\hat{c}$ deviations are measured as percentages

of a.)

For comparison, the tables also display the cost penalty incurred

by using band-limited flat spectrum inputs, i.e. those inputs for which
the design measure ξ satisfies

$$d\xi(\omega) = \pi\omega_b^{-1}d\omega \qquad\qquad \omega \in [0,\omega_b] \qquad\qquad (8)$$

$$= 0 \qquad\qquad \omega > \omega_b$$

for some ω_b. It follows, after some manipulation, that

$$\bar{M}_{aa} = \frac{b^2}{2a^2}\{(1-\frac{3c^2}{a^2})\frac{\tan^{-1}(a\omega_b)}{a\omega_b} + \frac{2c^2}{a^2} + (\frac{c^2}{a^2}-1)\frac{1}{1+a^2\omega_b^2}\} \qquad (9)$$

$$\bar{M}_{bb} = \frac{c^2}{a^2} + (1-\frac{c^2}{a^2})\frac{\tan^{-1}(a\omega_b)}{a\omega_b} \qquad\qquad (10)$$

from which the cost (A4) can be calculated.

It is easily shown that choosing ω_b as zero or infinite gives no
information, i.e. infinite cost. Two intermediate selections are made
here. One is the rather artificial choice $\omega_b = \omega_b^*$, the best band frequency
in the sense of minimizing the cost over the class of inputs specified by
(A8). The second choice is $\omega_b = 10\omega_b^*$, thus covering a sufficiently wide
spectrum of frequencies so as to include any area of a priori interest.

The following points arise:

(i) In all cases, the wide band ($10\omega_b^*$) input is far from optimal and it
preferable to use the optimal design based on the estimated parameter
values.

(ii) Inaccuracies in $\hat{c}$ have a less harmful effect than those in $\hat{a}$ when
a, b are being estimated.

(iii) When a/c is close to unity, the best flat spectrum input is nearly
optimal (Table 3) and this suggests that an input of this type may be

satisfactory rather than going for an optimal design.

 (iv) The case a = c (Table 2) is a degenerate case in which C(s) is

a factor of A(s). In general, if this occurs and the polynomial orders

satisfy the conditions

$$q = 0, \qquad m = n-r \tag{11}$$

as in the present example, then a hyperplane situation arises and an

optimal input is possible with continuous spectrum. In the present case,

the best flat-spectrum input is optimal. Point (iii) is again relevant.

     Points (iii) and (iv) suggest that where the process structure

approximately gives rise to the hyperplane conditions, then it may not

be advantageous to pursue the 'optimal' course (at least initially).

     Payne and Goodwin [P1] suggest that the problem of dealing with a

diffuse prior distribution for the parameter estimates can be tackled

bu choosing the design to minimize the *average* cost over the distribution.

Ng and Goodwin [N5] point out that this does not guard against a single

non-informative experiment and propose maximising the cost function

$$\prod_{i=1}^{n} \det \bar{M}(\theta_i)$$

where $\theta_1, \theta_2, \ldots, \theta_n$, the n values of the parameter p-vector θ, may be

chosen as the vertices of a hypercuboid enclosing the region in

parameter space having non-zero probability. Both approaches inject

robustness into the optimal designs that result.

$\hat{a}$: percentage deviations from a					
-20	-10	-1	+1	+10	+20
COST 7.2252	1.6384	0.0153	0.0151	1.4366	5.5145

$\hat{c}$: percentage deviations from c					
-20	-10	-1	+1	+10	+20
COST $.9502 \times 10^{-3}$	$.1650 \times 10^{-3}$	$.1232 \times 10^{-5}$	$.1168 \times 10^{-5}$	$.8992 \times 10^{-4}$	$.2784 \times 10^{-3}$

Flat Spectrum Inputs	
$\omega_b = \omega_b^*$	$\omega_b = 10\omega_b^*$
COST 43.6747	516.1305

$\omega_b^* = 4.2224$

$\omega^* = 1.7243$, $\quad \det \bar{M}(\omega^*)^{-1} = 0.9418 \times 10^{-3}$

Table 1: $a = 1$, $c = 10$

(Note: In each table cost is calculated as percentage increase on the optimum value.)

$\hat{a}$: percentage deviations from a					
-20	-10	-1	+1	+10	+20

COST	19.4265	4.4097	0.0404	0.0396	3.6129	13.0563

$\hat{c}$: percentage deviations from c					
-20	-10	-1	+1	+10	+20

COST	4.2390	1.0671	0.0101	0.0099	0.8794	2.9730

Flat Spectrum Inputs	
$\omega_b = \omega_b^*$	$\omega_b = 10\omega_b^*$
0	308.1697

COST

$\omega_b^* = 2.3311$

$\omega^* = 1.0000, \quad \det \bar{M}(\omega^*)^{-1} = 4.0000$

Table 2: a = 1, c = 1

$\hat{a}$: percentage deviations from a					
-20	-10	-1	+1	+10	+20
COST 21.2159	4.5070	0.0390	0.0378	3.2993	11.5472

$\hat{c}$: percentage deviations from c					
-20	-10	-1	+1	+10	+20
COST 3.1385	0.8589	0.0090	0.0090	0.8874	3.3166

Flat Spectrum Inputs	
$\omega_b = \omega_b^*$	$\omega_b = 10\omega_b^*$
COST 0.6329	312.8190

$\omega_b^* = 2.0896$

$\omega^* = 0.9020, \quad \det \bar{M}(\omega^*)^{-1} = 4.8312$

Table 3: a = 1, c = 0.9

5.A

$\hat{a}$: percentage deviations from a					
-20	-10	-1	+1	+10	+20
COST 1.9543	0.4244	0.0038	0.0037	0.3358	1.2156

$\hat{c}$: percentage deviations from a					
-20	-10	-1	+1	+10	+20
COST unstable region			$.3059 \times 10^{-6}$	0.0030	0.0490

Flat Spectrum Inputs	
$\omega_b = \omega_b^*$	$\omega_b = 10\omega_b^*$
COST 14.3858	2155.5854

$\omega_b^* = 1.0789$

$\omega^* = 0.5774,$ $\det \bar{M}(\omega^*)^{-1} = 9.4815$

Table 4: a = 1, c = 0

Chapter 6

SAMPLING RATE DESIGN

6.1   THE SAMPLING PROBLEM

The increasing use of digital computers in system identification
and control theory has emphasised the importance of discrete-time
models for natural processes that are, in essence, of a continuous
nature. This raises the important problem of the choice of sampling
times for accurate estimation of continuous-time system parameters
from sampled data corrupted by noise.

The problem of optimal sampling rate determination for linear system
identification has been discussed by Astrom [A5] and Gustavsson [G6].
To gain insight into the problem these authors considered the cases
where the system input is either absent or prespecified as discrete-
time white noise. In general, however, to achieve maximum return from
an experiment, a coupled design of the presampling filter, sampling
rate and test signal should be carried out. The general design problem,
with non-uniform sampling intervals and presampling filters, can be
formulated [G7][Z2][G9]. The general solution, however, is complex and
offers little insight.

Payne et al [P4] and Ng and Goodwin [N5] have developed a frequency
domain approach to the optimal coupled design of uniform sampling rate,
presampling filter and test signals. Following a brief exposition of
their method, the remainder of this chapter is concerned with the further
development of this approach within the Tchebycheff system framework.

6.2 RESUME

The framework developed in Chapter 5 for continuous-time systems can be used as a basis for analysis of the joint input design/sampling problem (Sections 3, 4).

Geometrical difficulties arise, in that $\hat{M}(\omega_c)$, the set of average information matrices per unit sample corresponding to normalised input power, is not necessarily either closed or convex. Nevertheless, it is shown that it is sufficient to consider only designs with discrete spectra whose design indices do not exceed p+2 (Section 5). A suitable design algorithm is then formulated (Section 6) and, finally, some properties of Φ-optimal designs are discussed, based on certain interlacing properties of canonical design measures (Section 7).

6.3 FREQUENCY DOMAIN APPROACH

Consider the continuous-time system described in Section (5.2)[.†] The average information matrix per unit time, denoted here by $\bar{M}_T$, is given by

$$\bar{M}_T = \text{Re} \int_0^{\omega_h} h(j\omega) h^*(j\omega) d\xi(\omega) \tag{1}$$

in the usual notation, where the band frequency $\omega_h (\leq \omega_c)$ is the lowest frequency for which the input power lies wholly in the band $[0,\omega_h]$, i.e.

$$\omega_h = \inf_{\omega \in [0,\omega_c]} \{\omega | \int_\omega^{\omega_c} d\xi(\omega) = 0\} \tag{2}$$

Consider the sampling scheme depicted in Figure 1.

---

†It is assumed throughout this chapter that the noise parameters are known.

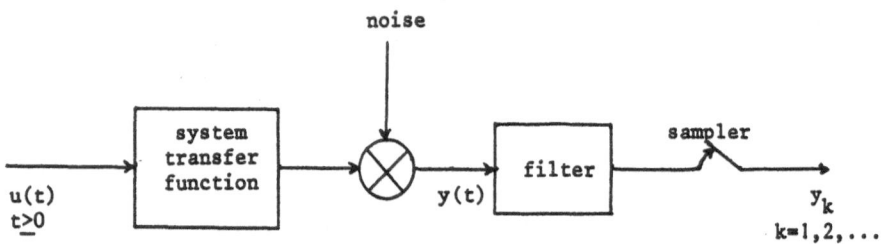

Figure 1

If the output y(t), t ε [0,T], is sampled at greater than the Nyquist

rate for ω_h, i.e. the sampling frequency ω_s satisfies the inequality

$$\omega_s > 2\omega_h \qquad\qquad (3)$$

then the sampler does not distort that part of the output spectrum arising

from the input. In general, however, that part of the output spectrum

due to the noise will be distorted due to aliasing unless a suitable

filter is used.

Aliasing leads to information loss in the sense that the matrix

$\bar{M}_T - \bar{M}_T^s$ is then positive definite where $\bar{M}_T^s$ denotes the average information

matrix per unit time after sampling. This loss of information can be

avoided by a suitable choice of presampling filter as follows:

Theorem 1 [P4]

If condition (3) holds, then any presampling filter with transfer

function F(s) satisfying

$$F(j\omega) = 0 \qquad \forall\ \omega \geq \omega_s/2$$

and invertible for $\omega\ \epsilon\ [0,\omega_h]$ is optimal in that it leads to the equality

$$\bar{M}_T^s = \bar{M}_T \qquad\qquad\qquad\qquad \#$$

In particular, the standard anti-aliasing filter [G2] is optimal
and has a transfer function F(s) given by

$$F(j\omega) = 1 \qquad \forall\ \omega \in [0,\omega_s/2)$$
$$\quad\ \ = 0 \qquad \text{otherwise}$$

6.4   THE DESIGN PROBLEM

The results of the last section can be used to develop a design
procedure for joint determination of the optimal input, presampling
filter and constant sampling rate for the case of constrained input
power and fixed total number of data samples.

The restriction on the number of samples is frequently met in
practice and arises from both the cost of data acquisition and limits
on computer storage. In this case it is reasonable to cost the
information per sample. The average information matrix per unit sample,
denoted by $\bar{M}$, is given by

$$\bar{M} = \frac{2\pi}{\omega_s}\,\bar{M}_T^s = \frac{2\pi}{\omega_s}\,\bar{M}_T \qquad\qquad\qquad (1)$$

using any filter and sampler for which Theorem (3.1) is valid.

To achieve Φ-optimality (as defined in Section (2.9)), it is
clear that (theoretically) the best choice of sampling rate is given
by

$$\omega_s = 2\omega_h \qquad\qquad\qquad\qquad (2a)$$

In practice, the optimal choice will be

$$\omega_s = 2(1+\epsilon)\omega_h \qquad (2b)$$

where ϵ (> 0) is related to the cut-off characteristics of the presampling filter. The factor $(1+\epsilon)$ appears only as a scaling factor, does not affect the input design spectrum and is therefore omitted below.

The choice (2a) yields a sampling interval Δ related to the input spectrum by

$$\Delta = 2\pi/\omega_s = \pi/\omega_h \qquad (3)$$

and leads to an (average) information matrix (per unit sample)

$$\bar{M} = \text{Re } \frac{\pi}{\omega_h} \int_0^{\omega_h} h(j\omega)\,h^*(j\omega)\,d\xi(\omega) \qquad (4)$$

With constrained input power, the design problem is to choose $\xi \in \Xi_1(\omega_c)$ such that $\Phi(\bar{M})$ is minimised. As before, ω_c is some cut-off frequency and $\omega_h \leq \omega_c$.

It follows immediately from the form of (4) that $\bar{M}$ is nonsingular if the input is persistently exciting in the sense of Theorem (5.4.1ii).

6.5   A GEOMETRIC APPROACH

The type of sampling scheme outlined in Sections 3 and 4 indicates that the geometric framework constructed in Chapter 5 for continuous-time systems may be of use here. The notation carries over from the last chapter.

Definition 1

Let $\hat{M}(\omega_c)$ denote the set of information matrices of the form (4.4) corresponding to the set $E_1(\omega_c)$ of design measures.

Note that, in contrast to previous cases, the relationship of design measure to information matrix, i.e. the mapping $\Xi_1(\omega_c) \to \hat{M}(\omega_c)$ is neither linear nor continuous (c f. (3.2.1), (5.3.5)).

Definition 2

Let $\hat{M}^{(p)}(\omega_c)$ denote the set in R^p isomorphic to $\hat{M}(\omega_c)$ such that $\underline{x} \in \hat{M}^{(p)}(\omega_c)$ if

$$\underline{x} = \frac{1}{\omega_h} \int_0^{\omega_h} \underline{v}(\omega) \, d\xi(\omega) \tag{1}$$

$$\xi \in \Xi_1(\omega_c), \qquad \omega_h \leq \omega_c$$

Consider the relationship of $\hat{M}^{(p)}(\omega_c)$ to the cone $M_c(\omega_c)$.

* Result 1

$$\hat{M}^{(p)}(\omega_c) \subset M_c(\omega_c)$$

Proof

Let $\underline{x} = \omega_h^{-1} \underline{x}' \in R^p$ where

$$\underline{x}' = \int_0^{\omega_h} \underline{v}(\omega) \, d\xi(\omega) \qquad \text{and} \qquad \xi \in \Xi_1(\omega_c)$$

Then

$$\underline{x}' \in M^{(p)}(\omega_c) \subset M_c(\omega_c)$$

But $\underline{x} \in \hat{M}^{(p)}(\omega_c)$ and lies on the positive ray through $\underline{x}'$. The result follows.   #

* Result 2

The set $\hat{M}^{(p)}(\omega_c)$ is not necessarily closed.

Proof

Consider the following counterexample for $p > 2$.

Define the sequence of points in R^p

$$\underline{x}(k) = \{(1-2^{-k})\underline{v}(\omega_1) + 2^{-k}\underline{v}(\omega_2)\}/\omega_2, \quad k = 0, 1, \ldots$$

where

$$0 < \omega_1 < \omega_2 < \omega_c$$

Each member of the sequence lies in $\hat{M}^{(p)}(\omega_c)$. The limit point $\underline{x}$ of the sequence is given by

$$\underline{x} = \underline{v}(\omega_1)/\omega_2$$

$$= \lambda v(\omega_1)/\omega_1 \quad \text{where} \quad \lambda = \omega_1/\omega_2 < 1$$

and therefore can be written in the form

$$\underline{x} = \frac{1}{\omega_1}\int_0^{\omega_1}\underline{v}(\omega)d\xi_1(\omega) \quad \xi_1 \notin \Xi_1(\omega_c)$$

For $p > 2$, the point $\underline{x}$ lies in Bd $M_c(\omega_c)$ and therefore ξ_1 is the unique design measure mapping into $\underline{x}$. Hence $\underline{x} \notin \hat{M}^{(p)}(\omega_c)$ and the result follows. #

The results below correct an erroneous theorem stated by Goodwin and Payne [G9, p. 22].

* Result 3

The set $\hat{M}^{(p)}(\omega_c)$ is not necessarily convex.

Proof

Consider the following counterexample for $p > 4$.

Let

$$\underline{x}_1, \underline{x}_2 \in \hat{M}^{(p)}(\omega_c)$$

where

$$\underline{x}_i = \underline{v}(\omega_i)/\omega_i \qquad i = 1, 2 \qquad\qquad (2)$$

$$0 < \omega_1 < \omega_2 < \omega_c$$

Consider the point

$$\underline{x} = (1-\alpha)\underline{x}_1 + \alpha\underline{x}_2 \qquad \alpha \in (0,1)$$

Then $\underline{x}$ can be expressed in the forms

$$\underline{x} = \frac{1}{\omega_2} \int_0^{\omega_2} \underline{v}(\omega)\,d\xi(\omega) = \frac{1}{\omega_2} \sum_{i=1}^{2} \lambda_i \underline{v}(\omega_i) \qquad\qquad (3)$$

where

$$\lambda_1 = (1-\alpha)\omega_2/\omega_1, \qquad \lambda_2 = \alpha$$

But $I(\omega_c;\xi) = 2 < p/2$ and therefore the points $\underline{x}_1$, $\underline{x}_2$, $\underline{x}$ all lie in Bd $M_c(\omega_c)$ and have unique representations. In particular, ξ is the unique design measure mapping into $\underline{x}$ and

$$\int_0^{\omega_2} d\xi(\omega) = \sum_{i=1}^{2} \lambda_i = (1-\alpha)\omega_2/\omega_1 + \alpha > 1$$

Hence there exists no design measure in $\Xi_1(\omega_c)$ for which (3) holds. Therefore $\underline{x} \notin \hat{M}^{(p)}(\omega_c)$ and the result follows.   #

Results 2 and 3 raise two problems:

(i) Does a Φ-optimum exist in $\hat{M}^{(p)}(\omega_c)$?

(ii) Can the search for a Φ-optimal design be restricted to the set of discrete design measures $\mathcal{D}_1(\omega_c)$?

The first problem is dealt with in the next section. The second problem arises because $\hat{M}^{(p)}(\omega_c)$ is not convex and therefore Caratheodory's theorem cannot be directly invoked(c f. Theorem (3.2.1)). This problem

is now resolved via the following representation theorem (c.f. Theorems

(3.2.1) and (5.4.2)).

* Theorem 1

Let $\bar{M}(\xi_1) \in \hat{M}(\omega_c)$ where $\xi_1 \in \Xi_1(\omega_c)$. Then there exists $\xi_2 \in \mathcal{D}_1(\omega_c)$

Such that $\bar{M}(\xi_1) = \bar{M}(\xi_2)$ and with design index satisfying the inequality

$$I(\omega_c;\xi_2) \leq p+2 \qquad (4)$$

If $M^{(p)}(\omega_c)$ is a hyperplane in R^p, then

$$I(\omega_c;\xi_2) \leq (p+1)/2 \qquad (5)$$

Proof

Let the design measure ξ_1 have band frequency ω_h and map into $\underline{x}_1$

in $\hat{M}^{(p)}(\omega_c)$, i.e.

$$\underline{x}_1 = \frac{1}{\omega_h}\int_0^{\omega_h} \underline{v}(\omega)d\xi_1(\omega) = \frac{1}{\omega_h}\underline{x}' \qquad (6)$$

where

$$\underline{x}' \in M^{(p)}(\omega_h) \subseteq M^{(p)}(\omega_c)$$

Note that the hyperplane conditions (Result (5.7.1)) are
independent of the cut-off frequency and therefore

$$M^{(p)}(\omega_h) \text{ hyperplane} \iff M^{(p)}(\omega_c) \text{ hyperplane}$$

Consider the following cases:

(i) $M^{(p)}(\omega_c)$ is not a hyperplane.

(a) $\underline{x}' \in \text{Bd } M^{(p)}(\omega_h)$. In this case ξ_1 must be discrete (Result 5.7.3))

and

$$I(\omega_c; \xi_1) \le p \qquad (7)$$

(b) $\underline{x}' \in \text{Int } M^{(p)}(\omega_h)$. The convexity of $M^{(p)}(\omega_h)$ ensures that there exists $\alpha \in (0,1)$, $\underline{x}_0 \in M^{(p)}(\omega_h)$ such that

$$\underline{x}' = (1-\alpha)\underline{x}_0 + \alpha \underline{v}(\omega_h) \qquad (8)$$

and, using Theorem (5.4.2), $\underline{x}_0$ can be expressed in the form

$$\underline{x}_0 = \sum_{i=1}^{p+1} \lambda_i' \underline{v}(\omega_i)$$

where

$$\lambda_i' \ge 0, \ i = 1,\ldots,p+1; \qquad \sum_{i=1}^{p+1} \lambda_i' = 1$$

$$0 \le \omega_1 < \ldots < \omega_{p+1} \le \omega_h$$

Therefore, from (8), $\underline{x}'$ can be expressed in the form

$$\underline{x}' = \sum_{i=1}^{p+2} \lambda_i \underline{v}(\omega_i) \qquad (9)$$

where

$$\lambda_i \ge 0, \ i = 1,\ldots,p+2; \qquad \sum_{i=1}^{p+2} \lambda_i = 1$$

$$0 \le \omega_1 < \ldots < \omega_{p+2} = \omega_h$$

Hence

$$\underline{x}_1 = \frac{1}{\omega_h} \underline{x}' = \frac{1}{\omega_h} \int_0^{\omega_h} \underline{v}(\omega)\,d\xi_2(\omega)$$

where $\xi_2 \in \mathcal{D}_1(\omega_h) \subseteq \mathcal{D}_1(\omega_c)$ and is given by (9) so that

$$I(\omega_c; \xi_2) \le p+2 \qquad (10)$$

(ii) $M^{(p)}(\omega_c)$ is a hyperplane.

(a) $\underline{x}' \in \text{Bd } M_c(\omega_h)$. It follows from result (5.7.2) that

$$I(\omega_c;\xi_1) \leq I(\omega_h;\xi_1) + \frac{1}{2} \leq (p+1)/2 \tag{11}$$

(b) $\underline{x}' \in \text{rint } M^{(p)}(\omega_h)$. Choose ξ_2 to be the upper principal design measure mapping to $\underline{x}'$. Then $\xi_2 \in \mathcal{D}_1(\omega_h)$ and $I(\omega_h;\xi_2) = p/2$. Finally

$$I(\omega_c;\xi_2) \leq I(\omega_h;\xi_2) + \frac{1}{2} = (p+1)/2 \tag{12}$$

This completes the proof.   #

Theorem 1 proves that only input designs with discrete spectra need to be considered in the search for a Φ-optimum.

6.6 FINDING THE Φ-OPTIMUM

Using Theorem (5.1), the optimal design problem can be posed in the following way:

Problem A

Find $\underline{x}_A \in R^p$ such that

$$\underline{x}_A = \arg \inf_{\underline{x} \in \hat{M}^{(p)}(\omega_c)} \Phi(\underline{x}) \tag{1a}$$

where $\underline{x} \in \hat{M}^{(p)}(\omega_c)$ if

$$\underline{x} = \frac{1}{\omega_h} \int_0^{\omega_h} \underline{v}(\omega) d\xi(\omega) = \frac{1}{\omega_h} \sum_{i=1}^{\ell} \lambda_i \underline{v}(\omega_i) \tag{1b}$$

$$\omega_h = \inf_{\omega \in [0,\omega_c]} \{\omega | \int_\omega^{\omega_c} d\xi(\omega) = 0\} \tag{1c}$$

$$0 \leq \omega_1 \leq \cdots \leq \omega_\ell \leq \omega_c; \qquad \ell = p+2 \qquad\qquad (1d)$$

$$\lambda_i \geq 0, \qquad i = 1, \ldots, \ell; \qquad \sum_{i=1}^{\ell} \lambda_i = 1 \qquad\qquad (1e)$$

This formulation brings out the discontinuous nature of the mapping $\mathcal{D}_1(\omega_c) \rightarrow \hat{M}^{(p)}(\omega_c)$. Assume that $\lambda_{\ell-1} > 0$. Then, in the limit as λ_ℓ is taken to zero, ω_h flips from ω_ℓ to $\omega_{\ell-1}$, giving rise, in general, to a jump in the value of the cost function and its derivatives. This behaviour indicates that it may be difficult to use a standard search algorithm to solve Problem A directly. This difficulty can be overcome by reformulating the problem.

First, given that $\hat{M}^{(p)}(\omega_c)$ is not in general closed, it is useful to establish the following result:

* Result 1

$$\underline{x}_A \in \hat{M}^{(p)}(\omega_c) \cap \mathrm{Bd}\, \hat{M}^{(p)}(\omega_c)$$

Proof

$$\underline{x}_A \in \mathrm{Bd}\, \hat{M}^{(p)}(\omega_c) \qquad\qquad (\text{see Section } (3.2))$$

Let $\chi(\omega_c)$ denote the closure of $\hat{M}^{(p)}(\omega_c)$. Then

$$\underline{x} \triangleq \frac{1}{\omega_\ell} \sum_{i=1}^{\ell} \lambda_i \underline{v}(\omega_i) \in \hat{M}^{(p)}(\omega_c) \qquad \text{if } \lambda_\ell > 0$$

$$\underline{x}_1 \triangleq \frac{1}{\omega_\ell} \sum_{i=1}^{\ell-1} \lambda_i \underline{v}(\omega_i) \in \chi(\omega_c)$$

subject to (1d) and (1e).

The Φ-optimum $\underline{x}_A$ is attained in $\chi(\omega_c)$. Assume that

$$\underline{x}_A = \frac{1}{\omega_\ell} \sum_{i=1}^{\ell-1} \lambda_i \underline{v}(\omega_i)$$

where $\omega_{\ell-1} < \omega_\ell$. Then

$$\underline{x}' \triangleq \frac{1}{\omega_{\ell-1}} \sum_{i=1}^{\ell-1} \lambda_i \underline{v}(\omega_i) \in \hat{M}^{(p)}(\omega_c)$$

and $\Phi(\underline{x}') < \Phi(\underline{x}_A)$ which contradicts the Φ-optimality of $\underline{x}_A$. Therefore $\underline{x}_A \in \hat{M}^{(p)}(\omega_c)$ and the result follows.   #

The proof of Result 1 indicates that the continuity difficulties inherent in the formulation of Problem A can be bypassed by considering the following alternative:

Problem B

Find $\underline{x}_B \in R^p$ such that:

$$\underline{x}_B = \arg \inf_{\underline{x} \in \chi(\omega_c)} \Phi(\underline{x}) \tag{2a}$$

where

$$\underline{x} = \frac{1}{\omega_\ell} \sum_{i=1}^{\ell} \lambda_i \underline{v}(\omega_i) \tag{2b}$$

subject to (1d) and (1e).

It is clear that $\underline{x}_A$ is of the form (2b) with $\lambda_\ell > 0$ and therefore $\underline{x}_B = \underline{x}_A$. In addition the mapping $\mathcal{D}_1(\omega_c) \to \chi(\omega_c)$ is continuous. Thus the computational solution to Problem B can be carried out without discontinuity problems.

It may be possible to attain Φ-optimality with less that p+2 frequencies so that the optimization dimension is reduced. Again, identifiability considerations demand that at least $[(p+1)/2]$ distinct frequencies are present in the input spectrum.

Example 1

Consider the system (p=3)

$$y(s) = \frac{1+s}{1+0.5s} u(s) + e(s)$$

The set $M^{(3)}(\infty)$ is a hyperplane in R^3 and det $\bar{M}_T^{-1}$ is minimized by the canonical design

Frequency	Power Proportion
0.123	0.337
3.482	0.663

leading to

$$\text{var } \hat{a}_1 = 4.00/T$$

$$\text{var } \hat{b}_0 = 3.00/T$$

$$\text{var } \hat{b}_1 = 12.75/T$$

where T is the experiment time.

On sampling (according to the scheme described in Section 3) there exists a D-optimal lower principal design given by

Frequency	Power Proportion
0	1/3
1.549	2/3

leading to an optimal sampling interval, Δ^*, given by

$$\Delta^* = \pi/1.549 = 2.028$$

and

$$\text{var } \hat{a}_1 = 4.734/N$$

$$\text{var } \hat{b}_0 = 1.479/N$$

$$\text{var } \hat{b}_1 = 9.443/N$$

where N is the total number of samples.

Note that the e exist an infinite number of canonical designs equivalent to (3) [Result (5.7.2)]. It is shown in the next section, however, that, on sampling, the D-optimal design (5) is unique.

6.7   PROPERTIES OF THE Φ-OPTIMUM

A sufficient condition for the Φ-optimum $\underline{x}^*$ to lie in the interior of the cone $M_c(\omega_c)$ can be derived as in Section (5.6). The condition is that

$$\lim_{\omega \to \infty} \omega^{-1} \underline{\dot{v}}(\omega) = 0$$

and this leads to the inequality

$$q-r \leq n-m \tag{1}$$

Note that equality is allowed in (1), (c.f. Result (5.6.3)).

If $\underline{x}^* \in Bd\ M_c(\omega_c)$, then it has a unique representation with index $(p-1)/2$ whose spectrum includes ω_c (Section (5.6)). Thus the Φ-optimal input design is unique with optimal sampling interval π/ω_c (i.e. sample as fast as possible).

Assume that the Φ-optimum $\underline{x}^*$ lies in the interior of the cone $M_c(\omega_c)$. Labelling the spectral frequencies of the upper and lower principal representations of $\underline{x}^*$ by upper and lower bars respectively, then Theorem (3.6.3) leads to the ordering

$$0 \leq \underline{\omega}_1 < \bar{\omega}_1 < \underline{\omega}_2 < \ldots < \bar{\omega}_{[p/2]} < \underline{\omega}_{[(p+1)/2]} < \bar{\omega}_{[p/2]+1} = \omega_c \tag{2}$$

Further, if ξ is any other design measure representing $\underline{x}^*$ and lying in $\Xi(\omega_c)$, then every non-empty open interval $(0,\omega_1)$, $(\underline{\omega}_{[(p+1)/2]},\omega_c)$,

$(\underline{\omega}_i, \underline{\omega}_{i+1})$ and $(\bar{\omega}_i, \bar{\omega}_{i+1})$ contains a point of increase of ξ (see Main Appendix). In particular, if ξ is discrete, then its band frequency lies in $(\underline{\omega}_{[(p+1)/2]}, \omega_c)$ and the optimal sampling interval Δ^* therefore has an upper bound given by

$$\Delta^* \leq \pi / \underline{\omega}_{[(p+1)/2]} \qquad (3)$$

In general, the Φ-optimal design is not unique. If $M^{(p)}(\omega_c)$ is a hyperplane, however, uniqueness follows. Let $\underline{x}^* = \underline{x}/\omega_1$ where ω_1 is the band frequency of some design measure ξ_1 representing $\underline{x}$ in the relative interior of $M^{(p)}(\omega_c)$. Then $\xi_1 \in \Xi_1(\omega_c)$ and must have the lowest band frequency of all design measures representing $\underline{x}$. Thus ξ_1 is the unique lower principal design and the upper bound (3) is achieved (see Example (6.1)).

Summing up:-

* Theorem 1

If the Φ-optimum $\underline{x}^*$ lies in the interior of the cone $M_c(\omega_c)$, then the optimal sampling interval Δ^* satisfies the inequality

$$\Delta^* \leq \pi / \omega_0 \qquad (4)$$

where ω_0 is the highest frequency in the lower principal representation of $\underline{x}^*$. If, in addition, $M^{(p)}(\omega_c)$ is a hyperplane in R^p, then $\Delta^* = \pi / \omega_0$ and the optimal input design is unique and equal to the lower principal representation of $\underline{x}^*$.

If $\underline{x}^* \in \text{Bd } M_c(\omega_c)$ then $\Delta^* = \pi / \omega_c$ and the optimal input design spectrum is unique with index $(p-1)/2$ and contains ω_c.

Remark

Note that, where D-optimal principal designs exist, the power proportions are known a priori. This follows from Section (4.2) with little modification.   #

The interlacing properties of design measures representing $\underline{x}^*$ emphasize the importance of principal representations in indicating both a suitable sampling rate and test signal spectrum. Unfortunately, it appears to be difficult finding a principal representation of $\underline{x}^*$ without first finding $\underline{x}^*$ itself.

If $\underline{x}^*$ is known then its principal representations can be generated by finding the design that achieves the global (zero) minimum of the Euclidean norm

$$\left|\left| \underline{x}^* - \sum_{i=1}^{\ell} \lambda_i \underline{v}(\omega_i) \right|\right|$$

where

$$\lambda_i \geq 0, \quad i = 1, \ldots, \ell$$

$$0 \leq \omega_1 < \ldots < \omega_\ell \leq \omega_c$$

and ℓ, ω_1, ω_ℓ can be suitably constrained to yield either the upper or lower principal design [c.f. Section (3.6)].

6.8   CONCLUDING REMARKS

This chapter has discussed a particular scheme for the joint determination of optimal test signal and constant sampling rate for system parameter identification and it has been shown that the optimization problem is at most $(2p+3)$-dimensional.

The filtering/sampling scheme adopted avoids the need for dealing with the messy technical manipulations that arise when continuous-time stochastic processes are sampled directly (see e.g. MacGregor [M6]). However, estimation accuracy for noise characteristics depends on the choice of sampling rate and this complication has been avoided by assuming known noise parameters.

Chapter 7

CONCLUSIONS AND FURTHER RESEARCH

In this work certain aspects of the experiment design problem for dynamic system identification have been investigated within a geometric framework drawing on the classical theory of Tchebycheff systems and their associated moment spaces.

In general it is concluded that the geometric approach is fruitful, giving a qualitative insight into the experiment design problem that is usually lacking, together with quantitative results that can lead to a substantial reduction in the dimensions of the design optimisation problem. It would be of interest to further investigate the possibility of establishing necessary and sufficient conditions for the existence of Φ-optimal canonical designs. Extending the theory to the case of single input, multiple output systems would appear to offer few problems, but the more interesting multiple input case necessitates the introduction of matrix design measures and the way forward is not so clear.

The sequential design approach developed in Chapter 4 is relatively expensive computationally and often slow to converge, except in the hyperplane case. Such algorithms, however, have the advantage of global convergence to a D-optimal design and, by employing frequency removal and round-off, they could be used to generate a relatively parsimonious design close to D-optimal and indicating the minimum number of frequencies necessary to attain the D-optimum. Such a design could then be used to initiate a direct search procedure when an accurate solution is demanded. The behaviour of sequential design algorithms with cost criteria other than determinant is of interest.

REFERENCES

[A1] AOKI M. and STALEY R.M., Some Computational Considerations in Input Signal Synthesis Problems, 2nd International Congress on Computing Methods, sponsored by SIAM, San Remo, Italy, September 1968.

[A2] ASTROM K.J. and EYKHOFF P., System Identification - A Survey, Automatica, 7, pp. 123-162, 1971.

[A3] AOKI M. and STALEY R.M., On Input Signal Synthesis in Parameter Idnetification, Automatica, 6, pp. 431-440, 1970.

[A4] ASTROM K.J., BOHLIN T. and WENSMARK S., Automatic Construction of Linear Stochastic Dynamic Models for Industrial Processes with Random Disturbances using Operating Records, Technical Paper 18.150, IBM Nordic Laboratory, 1965.

[A5] ASTROM K.J., On the Choice of Sampling Rates in Parameter Identification of Time Series, Information Sciences, 1, pp. 273-287, 1969.

[A6] ATWOOD C.L., Sequences Converging to D-optimal Designs of Experiments. Annals of Statistics, 1, pp. 342-352, 1973.

[A7] ASTROM K.J., Introduction to Stochastic Control Theory, Academic Press, New York, 1970.

[A8] AOKI M., Optimization of Stochastic Systems, Academic Press, New York, 1967.

[A9] ARIMOTO S. and KIMURA H., Optimal Input Test Signals for System Identification - An Information Theoretic Approach, Int. J. Systems Science, 1, 3, pp. 279-290, 1971.

[A10] AKHIEZER N.I. and GLAZMAN I.M., Theory of Linear Operators in Hilbert Space, Vol. 1, Frederick Ungar Pub. Co., New York, 1966.

[B1] BAEYENS R. and JAQUET B., Survey of Applications of Identification
      on Power Generation and Distribution, Proc. 3rd IFAC Symposium on
      Identification and System Parameter Estimation, The Hague, 1973.

[E1] EYKHOFF P., VAN DER GRITEN P.M., KWAKERNAAK H. and VELTMAN B.P.,
      Systems Modelling and Identification, Proc. 3rd IFAC Congress,
      London, 1966.

[E2] EYKHOFF P., Process Parameter and State Estimation, Automatica, 4,
      pp. 205-233, 1968.

[E3] EYKHOFF P., System Identification, Wiley, London, 1974.

[F1] FEDOROV V.V., Theory of Optimal Experiments, Academic Press,
      New York, 1972.

[F2] FEDOROV V.V., Sequential Methods for Design of Experiments in the
      Study of the Mechanism of a Phenomenon, New Ideas in Experiment
      Design, Ed. by V.V. Nalimov, Moscow, 1969.

[F3] FEDOROV V.V., The Design of Experiments in the Multiresponse Case,
      Theory of Probability and Its Applications, 16, 2, pp. 323-332, 1971.

[G1] GRAUPE D., Identification of Systems, Van Nostrand Reinhold,
      New York, 1972

[G2] GUSTAVSSON I., Survey of Applications of Identification in Chemical
      and Physical Processes, Automatica, 11, pp. 3-24, 1975.

[G3] GOODWIN G.C., Input Synthesis for Minimum Covariance State and
      Parameter Estimation, Electronics Letters, 5, 21, 16th October 1969.

[G4] GOODWIN G.C. and PAYNE R.L., Design and Characterization of Optimal
      Test Signals for Linear SISO Parameter Estimation, Paper TT-1, 3rd
      IFAC Symposium, The Hague, 1973.

[G5] GUPTA N.K., MEHRA R.K. and HALL W.E. Jnr., Application of Optimal
Input Synthesis to Aircraft Parameter Estimation, ASME J. Dynamic
System Measurement, 98, 2, pp. 139-145, 1976.

[G6] GUSTAVSSON I., Choice of Sampling Intervals for Parametric Identification,
Div. of Aut. Contr., Lund Inst. of Technology, Report 7103, 1971.

[G7] GOODWIN G.C., ZARROP M.B. and PAYNE R.L., Coupled Design of Test
Signal, Sampling Intervals and Filters for System Identification,
IEEE Trans AC-19, 6, pp. 748-752, 1974.

[G8] GOODWIN G.C., MURDOCH J.C. and PAYNE R.L., Optimal Test Signal Design
for Linear SISO System Identification, Int. J. Control, 17, 1,
pp. 45-55, 1973.

[G9] GOODWIN G.C. and PAYNE R.L., Choice of Sampling Intervals, Tech.
Report, EE7417, Dept. of Elec. Eng. Univ. of Newcastle, NSW, 1974.

[H1] HANNAN E.J., Time Series Analysis, Methuen, 1967.
[H2] HADLEY G., Linear Algebra, Addison-Wesley, 1969.

[J1] JAVAHERIAN H., Optimal Input Experiment Design for Parameter
Estimation, MSc Dissertation, Department of Computing and Control,
Imperial College, London, 1974.

[K1] KIEFER J., Optimal Experimental Designs, J. Royal Stat. Soc. B21,
pp. 272-319, 1959.

[K2] KIEFER J. and WOLFOWITZ J., Optimum Designs in Regression Problems,
Ann. Math. Stat., 30, pp. 271-294, 1959.

[K3] KIEFER J. and WOLFOWITZ J., The Equivalence of Two Extremum Problems,
Canadian J. Math., 12, pp. 363-366, 1960.

[K4] KIEFER J., Optimum Designs in Regression Problems, II, Ann. Math.
Stat., 32, pp. 298-325, 1961.

[K5]   KIEFER J., Two More Criteria Equivalent to D-optimality of Designs, Ann. Math. Stat., 33, pp. 792-796, 1962.

[K6]   KIEFER J., General Equivalence Theory for Optimum Designs (Approximate Theory), Ann. Math. Stat., 2, pp. 849-879, 1974.

[K7]   KARLIN S. and STUDDEN W.J., Optimal Experimental Designs, Ann. Math. Stat., 37, pp. 783-815, 1966.

[K8]   KARLIN S. and STUDDEN W.J., Tchebycheff Systems with Applications to Analysis and Statistics, Wiley-Interscience, 1966.

[L1]   LOPEZ-TOLEDO A.A., Optimal Inputs for Identification of Stochastic Systems, PhD Thesis, MIT, 1974.

[L2]   LEVIN M.J., Optimal Estimation of Impulse Response in the Presence of Noise, IRE Trans. Circuit Theory, CT-7, pp. 50-56, 1960.

[L3]   LEVADI V.S., Design of Input Signals for Parameter Estimation, IEEE Trans. AC-11, 2, pp. 205-211, 1966.

[L4]   LJUNG L., Chacterization of the Concept of 'Persistently Exciting' in the Frequency Domain, Div. of Automatic Control, Lund Inst. of Technology, Report 7119, 1971.

[L5]   LINDLEY D.V., On a Measure of the Information Provided by an Experiment, Ann. Math. Stat., 27, pp. 986-1005, 1956.

[L6]   LUENBERGER D.G., Optimization by Vector Space Methods, Wiley, 1968.

[M1]   MEHRA R.K., Optimal Inputs for Linear System Identification, IEEE Trans. AC-19, 3, pp. 192-200, 1974.

[M2]   MEHRA R.K., Optimal Input Signals for Parameter Estimation in Dynamic Systems - A Survey and New Results, IEEE Trans. AC-19, 6, pp. 753-768, 1974.

[M3] MEHRA R.K., Frequency Domain Synthesis of Optimal Inputs for

      Parameter Estimation, Div. of Eng. and Appl. Physics, Harvard Univ.,

      Tech. Report 645, 1973.

[M4] MEHRA R.K., Synthesis of Optimal Inputs for MIMO Systems with

      Process Noise, Div. of Eng. and Appl. Physics, Harvard Univ.,

      Tech. Report 649, 1974.

[M5] MÜLLER P.C. and WEBER H.I., Analysis and Optimization of Certain

      Quantities of Controllability and Observability for Linear Dynamical

      Systems, Automatica, 8, pp. 237-246, 1972.

[M6] MACGREGOR J.F., Optimal Choice of the Sampling Interval for Discrete

      Process Control, Technometrics, 18, 2, pp. 151-160, 1976.

[N1] NIEMAN R.E., FISHER D.G. and SEBORG D.E., A Review of Process

      Identification and Parameter Estimation Techniques, Int. J. Control,

      13, pp. 209-264, 1971.

[N2] NAHI N.E. and WALLIS D.E. Jr., Optimal Inputs for Parameter

      Estimation in Dynamic Systems with White Observation Noise, Paper

      IV-A5, Proc. JACC, Boulder, Colorado, pp. 506-512, 1969.

[N3] NAHI N.E. and NAPJUS G.A., Design of Optimal Probing Signals for

      Vector Parameter Estimation, Paper W9-5, Preprints IEEE Conf. on

      Decision and Control, Miami, Florida, 1971.

[N4] NAPJUS G.A., Design of Optimal Inputs for Parameter Estimation,

      PhD Dissertation, USC, 1971.

[N5] NG T.S. and GOODWIN G.C., On Optimal Choice of Sampling Strategies

      for Linear System Identification, Int. J. Control, 23, pp. 459-475,

      1976.

[P1] PAYNE R.L. and GOODWIN G.C., A Bayesian Approach to Experiment
      Design with Applications to Linear Multivariable Dynamic Systems,
      IMA Conference on Computational Problems in Statistics, Univ. of
      Essex, 1973.

[P2] PAYNE R.L., Optimal Experiment Design for Dynamic System
      Identification, PhD Thesis, Imperial College, London, 1974.

[P3] PAYNE R.L. and GOODWIN G.C., Simplification of Frequency Domain
      Experiment Design for SISO Systems, Publication 74/3, Dept. of
      Computing and Control, Imperial College, London, 1974.

[P4] PAYNE R.L., GOODWIN G.C. and ZARROP M.B., Frequency Domain Approach
      for Designing Sampling Rates for System Identification, Automatica,
      11, pp. 189-191, 1975.

[P5] PAYNE R.L., An A-Priori Estimate for the Information Matrix of a
      SISO Linear System, Publication 72/4, Dept. of Computing and
      Control, Imperial College, London, 1972

[P6] PAPOULIS A., Probability, Random Variables and Stochastic Processes,
      McGraw-Hill, New York, 1965.

[R1] RAULT A., Survey of Applications of Identification in Aeronautics
      and Space Vehicles, Proc. 3rd IFAC Symposium on Identification and
      System Parameter Estimation, The Hague, 1973.

[R2] RAJBMAN N.S., Survey of Applications of Identification Methods in
      the USSR, Proc. 3rd IFAC Symposium on Identification and System
      Parameter Estimation, The Hague, 1973.

[R3] REID D.B., Optimal Inputs for System Identification, Stanford Univ.
      Report No. SUDAAR 440, 1972.

[R4] ROTHENBERG T.J., Identification in Parametric Models, Econometrica,
      39, pp. 577-591, 1971.

[S1] ST JOHN R.C. and DRAPER N.R., D-optimality for Regression Designs, Technometrics, 17, 1, pp. 15-23, 1975.

[S2] SILVEY S.D., Statistical Inference, Penguin, 1970.

[S3] STALEY R.M. and YUE P.C., On System Parameter Identifiability, Information Sciences, 2, pp. 127-138, 1970.

[S4] ST JOHN R.C., Models and Designs for Experiments with Mixtures, PhD Thesis, Dept. of Statistics, Univ. of Wisconsin, 1973.

[S5] SCHWEPPE F.C., Uncertain Dynamic Systems, Prentice Hall, 1973.

[T1] TSE E., Information Matrix and Local Identifiability of Parameters, Paper 20-3, JACC, Colombus, Ohio, 1973.

[V1] VIORT B., D-optimal Designs for Dynamic Models: Part I Theory, Dept. of Statistics, Univ. of Wisconsin, Tech. Report 314, 1972.

[V2] VAN DEN BOS A., Selection of Periodic Test Signals for Estimation of Linear System Dynamics, Paper TT-3, Preprints of 3rd IFAC Symposium, The Hague, 1973.

[V3] VAN DEN BOS A., Estimation of Parameters in Linear Systems using Periodic Test Signals, Report, Cooperative Centre for Meas. and Control, Delft Univ. of Technology, Netherlands, 1973.

[V4] VAN DEN BOS A., Construction of Binary Multifrequency Test Signals, 1st IFAC Symposium on Identification, Prague, 1967.

[V5] VAN TREES H.L., Detection, Estimation and Modulation Theory, Part III, Wiley, 1971.

[W1] WHITTLE P., Some General Points in the Theory of Optimal Experiment Design, J. Royal Stat. Soc., B35, pp. 123-130, 1973.

[W2] WYNN H.P., The Sequential Generation of D-optimum Experimental Designs, Ann. Math. Stat., 41, pp. 1655-1664, 1970.

[W3] WHITE L.V., An Extension of the General Equivalence Theorem to Nonlinear Models, Biometrika, 60, pp. 345-348, 1973.

[Z1] ZARROP M.B. and GOODWIN G.C., Comments on 'Optimal Inputs for Linear System Identification', IEEE Trans. AC-20, 2, pp. 299-300, 1975.

[Z2] ZARROP M.B., Experiment Design for System Identification, MSc Dissertation, Dept. of Computing and Control, Imperial College, London, 1973.

[Z3] ZARROP M.B., GOODWIN G.C. and PAYNE R.L., Analytic Design of Optimal Input Spectra for Identification,Publication 74/25, Dept. of Computing and Control, Imperial College, London, 1974.

Main Appendix

TCHEBYCHEFF SYSTEMS

A1.   INTRODUCTION

In this appendix some basic results in the theory of Tchebycheff

systems and their associated moment spaces are brought together for

ease of reference. Proofs of results have been included only where

they add to ease of comprehension without excessive technical manipulation.

The source for the material is the standard work on Tchebycheff

system theory by Karlin and Studden [K8] and the notation used in this

appendix is theirs.

A2.   TCHEBYCHEFF SYSTEMS ON A CLOSED INTERVAL

Let u_0, u_1, $\ldots$, u_n denote continuous real-valued functions

defined on a closed finite interval [a,b]. These functions will be

called a *Tchebycheff system* over [a,b] (or T-*system*) provided that n+1st

order determinants

$$U \begin{pmatrix} 0,1,\ldots,n \\ t_0,t_1,\ldots,t_n \end{pmatrix} = \begin{vmatrix} u_0(t_0) & u_0(t_1) & \cdots & u_0(t_n) \\ u_1(t_0) & u_1(t_1) & \cdots & u_1(t_n) \\ \vdots & \vdots & & \vdots \\ u_n(t_0) & u_n(t_1) & \cdots & u_n(t_n) \end{vmatrix}$$

are strictly positive whenever $a \leq t_0 < t_1 < \ldots < t_n \leq b$.

The functions u_0, u_1, $\ldots$, u_n are referred to as a *complete

Tchebycheff system* (or CT-*system*) if $\{u_0, u_1, \ldots, u_r\}$ is a T-system for

each r = 0, 1, $\ldots$, n.

A function of the form $u = \sum_{i=0}^{n} a_i u_i$ where a_i are real numbers will

be called a (u-) *polynomial* and is said to be nontrivial if $\sum_{i=0}^{n} a_i^2 > 0$.

Note that, if $\{u_i\}_0^n$ is a T-system, a nontrivial polynomial possessing

n prescribed zeros $t_0 < t_1 < \ldots < t_{n-1}$ is constructed by the explicit

formula

$$u(t) = U \begin{pmatrix} 0,1,\ldots,n-1,n \\ t_0,t_1,\ldots,t_{n-1},t \end{pmatrix}$$

The polynomial u changes sign as t passes through each $t_i \in (a,b)$.

Definition 1

For any continuous function f on $[a,b]$ an isolated zero $t_0 \in (a,b)$

of f is called a *nonnodal zero* provided the function does not change

sign at t_0. All other zeros including zeros at the end points a and b

are called *nodal zeros*. The number of zeros of f is designated by $\tilde{z}(f)$

where nonnodal zeros are counted twice.

In Figure 1, the points t_1 and t_3 are nonnodal zeros while t_2 and

the end point a are nodal zeros. Therefore, in this case, $\tilde{z}(f) = 6$.

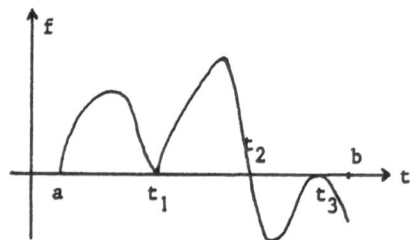

Figure 1

Let $\{\bar{u}_i\}_0^n$ denote the function sequence $\{u_0,u_1,\ldots,u_{n-1},-u_n\}$. The

following theorem embodies an alternative polynomial formulation of

T-systems.

Theorem 1

If $\{u_i\}_0^n$ is a T-system then $\tilde{z}(u) \le n$ for every nontrivial polynomial

u. Conversely, if $\{u_i\}_0^n$ is a system of continuous functions on $[a,b]$ and

$\tilde{z}(u) \leq n$ for every nontrivial polynomial u, then either $\{u_i\}_0^n$ or $\{u_i^-\}_0^n$ is a T-system.   #

Consider the problem of constructing nonnegative polynomials exhibiting a prescribed set of zeros.

Let $T = \{t_1,\ldots,t_k\}$ be an increasing set of distinct points in [a,b] and assign to each $t_i \in T$ a weight $\omega(t_i)$ defined by

$$\omega(t_i) = 2 \qquad t_i \in (a,b) \tag{1}$$

$$= 1 \qquad t_i = a \text{ or } b$$

Theorem 2

If $\{u_i\}_0^n$ is a T-system and $\sum_{i=1}^{k} \omega(t_i) \leq n$ then there exists a

nontrivial, nonnegative polynomial u vanishing precisely at the points of T. The only exception is that if n is even and exactly one of the end points a or b is in T then u(t) may vanish at the other end point as well.

If either of the following further conditions hold then without exception the polynomial may be constructed to vanish precisely on T:

(a) $\{u_i\}_0^{n-1}$ is a T-system

(b) $\{u_i\}_0^n$ is a T-system on [a',b'] where a' < a < b < b'.   #

Note that the T-system of interest in Chapter 3 satisfies condition (a) [see Result (3.4.1)].

The proof of Theorem 2 is technically messy [K8, pp. 28-30] and is omitted here. The following theorem can be proved in a similar way.

Theorem 3

Let $\{u_i\}_0^n$ be a T-system. Consider the set $T = \{t_1, \ldots, t_k\}$ with an assigned weight for each t_i as in (1), and suppose $\sum_{i=1}^{k} \omega(t_i) \leq n$. Then there exists a polynomial $u(t)$ such that $u(t) \neq 0$ for $t \in (a,b) - T$ and such that t_i is a nodal or nonnodal zero according as $\omega(t_i) = 1$ or 2, respectively.

Furthermore, under conditions (a) or (b) in Theorem 2 the above polynomial vanishes precisely on the set T.

A3.   MOMENT SPACES INDUCED BY T-SYSTEMS

The *moment space* M_{n+1} *with respect to the T-system* $\{u_i\}_0^n$ is generated in the following manner:

$$M_{n+1} = \{\underline{c} = (c_0, c_1, \ldots, c_n) \in E^{n+1} \mid c_i = \int_a^b u_i(t)\, d\sigma(t), \; i=0,1,\ldots,n\}$$

where the *measure* $\sigma(t)$ traverses the set of all nondecreasing right continuous functions of bounded variation.

Let C_{n+1} be the curve in E^{n+1} represented in parametric form as follows:

$$C_{n+1} = \{\gamma_t = (u_0(t), u_1(t), \ldots, u_n(t)) \mid a \leq t \leq b\}$$

Theorem 1

The moment space M_{n+1} is

(i)   a closed convex cone

(ii)   the convex conical hull of C_{n+1}.

Definition 1

The *index* $I(\underline{c})$ *of a point* $\underline{c}$ *in* M_{n+1} is defined to be the minimal number of points of C_{n+1} that are used in a convex representation of $\underline{c}$ under the convention that γ_a and γ_b are counted as half points while γ_t receives a full count for $t \in (a,b)$.

This leads to the basic representation theorems for points in either the boundary or interior of the cone M_{n+1}.

Theorem 2

A vector $\underline{c}^0 \in M_{n+1}$ $(\underline{c}^0 \neq 0)$ is a boundary point of M_{n+1} if and only if $I(\underline{c}^0) < (n+1)/2$. Moreover, every boundary point $\underline{c}^0$ admits a unique representation

$$c_i^0 = \sum_{j=1}^{k} \lambda_j u_i(t_j) \qquad i = 0,1,\ldots,n \qquad (1)$$

where $k \leq (n+2)/2$ and $\lambda_j > 0$, $j = 1,2,\ldots,k$.

Proof

Suppose $\underline{c}^0 \in \mathrm{Bd}\, M_{n+1}$ $(\underline{c}^0 \neq 0)$. Then there exists a supporting hyperplane to M_{n+1} at $\underline{c}^0$, i.e. there exist real constants $\{a_i\}_0^n$ such that

$$\sum_{i=0}^{n} a_i c_i \geq 0 \qquad \text{for all } \underline{c} \in M_{n+1}$$

$$\sum_{i=0}^{n} a_i c_i^0 = 0 \qquad (2)$$

Define $u^0(t) = \sum_{i=0}^{n} a_i u_i(t)$. Then, using (2), $u^0(t) \geq 0$ for $a \leq t \leq b$ and

$$\int_a^b u^0(t)\, d\sigma^0(t) = 0 \qquad \text{where} \quad c_i^0 = \int_a^b u_i(t)\, d\sigma^0(t)$$

Therefore, $u^0(t)$ vanishes at every point of increase of σ^0, i.e. the *spectrum* of σ^0 is confined to the zero set of $u^0(t)$. Thus, since $u^0(t) \geq 0$, it follows that:

$$2I(\underline{c}^0) = \tilde{z}(u^0) \leq n$$

using Theorem (A2.1), so that $I(\underline{c}^0) < (n+1)/2$.

Next, consider any representation of $\underline{c}^0$ of the form (1) where $T = \{t_j\}$ are the zeros of $u^0(t)$ augmented, if necessary, so that T contains n+1 distinct points. Since the n+1st order matrix whose general element is $u_i(t_j)$ has a nonzero (positive) determinant, then the coefficients λ_j in (1) are uniquely determined by $\underline{c}^0$.

Conversely, let $\underline{c}^0$ denote a vector of M_{n+1} for which $I(\underline{c}^0) < (n+1)/2$. Thus, $\underline{c}^0$ admits a representation of form (1). Invoking Theorem (A2.2) a nontrivial, nonnegative polynomial $u^0(t)$ can be constructed vanishing on $T = \{t_j\}$ corresponding to the representation. The coefficients of the polynomial $u^0(t)$ determine a supporting hyperplane to M_{n+1} at $\underline{c}^0$ and therefore $\underline{c}^0 \in \text{Bd } M_{n+1}$. #

Definition 2

The *index of a representation* of the form (1) is the *index of the set* $T = \{t_1, t_2, \ldots, t_k\}$ defined as that number obtained by counting interior points as one and the end points a and b as one half. The *index of a measure* σ generating $\underline{c}^0$ is taken as the index of the set T in the representation (1).

Definition 3

A *section* S of M_{n+1} is the intersection of a linear variety with the cone, with the property that $\lambda\underline{c}^0 \in S$ for a unique $\lambda > 0$ if $\underline{c}^0 \in M_{n+1}$ ($\underline{c}^0 \neq 0$).

Theorem 3

Let $\underline{c}^0 \in \text{Int } M_{n+1}$. For each t^* in $[a,b]$ there exists a representation

$$c_i^0 = \sum_{j=1}^{k} \lambda_j u_i(t_j), \qquad i = 0,1,\ldots,n$$

$$\lambda_j > 0, \qquad j = 1,2,\ldots,k$$

of index $(n+1)/2$ or $(n+2)/2$ which includes the point t^*.

Proof

Consider a section S of M_{n+1} such that $\underline{c}^0 \in \text{rint } S$. Choose λ so that $\underline{c}^* = (\lambda u_0(t^*),\ldots,\lambda u_n(t^*))$ lies in S and draw the line L from $\underline{c}^*$ through $\underline{c}^0$ to pierce the boundary of M_{n+1} in a second point $\underline{\tilde{c}}$. Clearly

$$\underline{c}^0 = \alpha \underline{\tilde{c}} + (1-\alpha)\underline{c}^* \qquad \text{for some } \alpha \in (0,1)$$

and $I(\underline{\tilde{c}}) = (n-1)/2$ or $n/2$, otherwise $I(\underline{c}^0) < (n+1)/2$ so that $\underline{c}^0 \in \text{Bd } M_{n+1}$, contrary to hypothesis. This leads to the required representation. #

A4.   INTERLACING PROPERTIES OF REPRESENTATIONS

Definition 1

Let $\underline{c}^0 \in \text{Int } M_{n+1}$. A representation for $\underline{c}^0$ of index $(n+1)/2$ is called *principal* and any representation of index not exceeding $(n+2)/2$ is called *canonical*. A canonical or principal representation is further designated by the term *upper* if it includes the end point b and the term *lower* otherwise.

Theorem (A3.3) asserts that for $\underline{c}^0 \in \text{Int } M_{n+1}$ there exists a canonical representation including any preassigned t^*. It can be shown [K9, pp. 45–49] that exactly two of these representations are principal, one upper and one lower.

For example, if n is even, then the construction used in the proof of Theorem (A3.3) leads to a lower (upper) principal representation of $\underline{c}^0$ if t* = a (=b).

The following interlacing properties are of interest, but only the first is proved here.

Lemma

Let $\underline{c}^0 \in \text{Int } M_{n+1}$ and let σ and $\sigma*$ be distinct measures representing $\underline{c}^0$, where σ is canonical. Then for every pair of interior points of increase t_j, t_{j+1} of σ there exists a point of increase of $\sigma*$ in the open interval (t_j, t_{j+1}). If σ is principal, the result remains true if t_j = a or t_{j+1} = b.

Proof

Let t_j, t_{j+1} denote the two consecutive points of increase of σ. If σ has index (n+2)/2 assume that the points lie in (a,b), while if σ is principal the possibility is allowed for that one of them is an end point.

If $\sigma*$ has no point of increase in (t_j, t_{j+1}), a polynomial u(t) can be constructed [Theorem (A2.3)] so that

$$u(t) \geq 0 \qquad t \in [t_j, t_{j+1}]$$

$$< 0 \qquad t \in [t_j, t_{j+1}]$$

vanishing on (a,b) precisely on the spectrum of σ. Note that t_j and t_{j+1} are nodal roots. Clearly

$$0 = \int_a^b u(t)[d\sigma*(t) - d\sigma(t)] = \int_a^b u(t)d\sigma*(t)$$

$$= \int_{[a,t_j]} u(t)d\sigma*(t) + \int_{[t_{j+1},b]} u(t)d\sigma*(t) \geq 0$$

and this is possible only if the spectrum of σ^* is part of the spectrum

of σ. However, σ has less than n+1 points of increase which implies

that σ and σ^* are not distinct. This contradiction leads to the required

result.   #

Result 1

For each $\underline{c}^0 \epsilon$ Int M_{n+1} there exist precisely two principal

representations. The roots of these representations strictly interlace.

Theorem 1

Let $\underline{c}^0 \epsilon$ Int M_{n+1} and consider two different canonical representations

σ_1, σ_2 of $\underline{c}^0$. Then the points of increase of σ_1, σ_2 lying in (a,b) strictly

interlace but they may possibly share one or both of the end points a or

b.

Corollary

Let $\underline{c}^0 \epsilon$ Int M_{n+1}. For any t* in (a,b), there exists a *unique*

canonical representation of $\underline{c}^0$ including t*.

Lecture Notes in Economics and Mathematical Systems

For information about Vols. 1–114 please contact your bookseller or Springer-Verlag

Lecture Notes in Economics and Mathematical Sy...